Experimental Design

Mike Nguyen

Experimental Design

Springer

Mike Nguyen
Data Sciences and Operations
University of Southern California
Los Angeles, CA, USA

ISBN 978-3-032-01838-0 ISBN 978-3-032-01839-7 (eBook)
https://doi.org/10.1007/978-3-032-01839-7

Mathematics Subject Classification: 62P20, 62Hxx, 62Kxx, 62J12, 62P25

This Springer imprint is published by the registered company Springer Nature Switzerland AG
The registered company address is: Gewerbestrasse 11, 6330 Cham, Switzerland

Contents

1 Introduction 1

2 Causal Inference 7
- 2.1 The Formal Notation of Causality 9
- 2.2 Simpson's Paradox 10
- 2.3 Experimental Versus Quasi-Experimental Designs 17
- 2.4 Hierarchical Ordering of Causal Tools 20
- 2.5 Types of Validity in Research 21
- 2.6 Types of Subjects in a Treatment Setting 26
- 2.7 Types of Treatment Effects 28

3 Experimental Design 49
- 3.1 Principles of Experimental Design 49
- 3.2 The Gold Standard: Randomized Controlled Trials 50
- 3.3 Selection Problem 51
- 3.4 Classical Experimental Designs 54
- 3.5 Advanced Experimental Designs 63
- 3.6 Emerging Research 73

4 Sampling 81
- 4.1 Population and Sample 81
- 4.2 Probability Sampling 82
- 4.3 Non-Probability Sampling 87
- 4.4 Unequal Probability Sampling 88
- 4.5 Balanced Sampling 89
- 4.6 Sample Size Determination 91

5 Analysis of Variance 93
- 5.1 Completely Randomized Design 94
- 5.2 Nonparametric ANOVA 143
- 5.3 Randomized Block Designs 147
- 5.4 Nested Designs 155

5.5 Sample Size Planning for ANOVA 161
5.6 Single-Factor Covariance Model 164

6 Multivariate Methods 169
6.1 Basic Understanding 170
6.2 Multivariate Analysis of Variance 213
6.3 Statistical Test Selection for Comparing Means 237

Bibliography 239

Chapter 1
Introduction

In the architecture of statistical reasoning, no structure is more foundational—or more often misunderstood—than experimental design. While regression analysis can uncover patterns and causal inference can suggest directionality, it is only through careful design that we gain the power to assert cause with confidence. This book, the fourth in the series, turns fully toward that power.

Experimental design is where ideas become interventions. It is where hypotheses are translated into treatments, where populations are sampled with intent, and where the rigor of the scientific method is most tangible. In this volume, we move from modeling relationships to constructing them, from observing systems to shaping them under controlled conditions.

This shift is not merely technical; it is epistemological. A well-designed experiment doesn't just provide cleaner data—it fundamentally changes the kinds of questions we are allowed to ask and the kinds of answers we are able to defend. In business, this might mean testing a new pricing strategy or optimizing a digital ad campaign. In policy, it could mean evaluating the rollout of a training program or subsidy. In both, the principles are the same: clarity in comparison, control over variation, and fidelity to the logic of inference.

We begin with the anatomy of experimental thinking: treatments, control groups, randomization, and replication. From there, we explore how design choices affect bias, precision, and generalizability. Sampling methods are addressed not as an afterthought, but as a key determinant of both internal and external validity. Analysis of variance (ANOVA), often treated as a statistical relic, is reintroduced here as a living tool—ideal for understanding the structure of designed experiments, particularly in factorial and blocked settings.

The volume culminates in multivariate methods tailored to experimental data, including MANOVA and multivariate regression for designs with multiple outcomes. Throughout, we emphasize how experimental tools are used not just in academic labs but in fast-paced business environments, where A/B testing and randomized field experiments now drive product decisions and strategic investments.

M. Nguyen, *Experimental Design*, https://doi.org/10.1007/978-3-032-01839-7_1

Code examples in R accompany each chapter, designed to make every design principle operational and transparent. But our aim is not just execution—it is reasoning. Readers are encouraged to think like designers: to anticipate threats to validity, to weigh trade-offs, and to see data not only as something collected but as something created through deliberate structure.

In the end, experimental design is not simply a methodology—it is a discipline of thought. It teaches us to be humble about inference, precise in our claims, and ambitious in the kinds of questions we are willing to test. For analysts, researchers, and decision-makers alike, that discipline is indispensable.

This book is dedicated to that discipline—and to the clarity, creativity, and care that it demands.

Code Replication

This book was built with R version 4.4.3 (2025-02-28 ucrt) and the following packages:

package	version	source
abind	1.4-8	CRAN (R 4.4.1)
agridat	1.24	CRAN (R 4.4.3)
ape	5.8-1	CRAN (R 4.4.3)
assertthat	0.2.1	CRAN (R 4.4.3)
backports	1.5.0	CRAN (R 4.4.0)
bookdown	0.43	CRAN (R 4.4.3)
boot	1.3-31	CRAN (R 4.4.3)
broom	1.0.8	CRAN (R 4.4.3)
bslib	0.9.0	CRAN (R 4.4.2)
cachem	1.1.0	CRAN (R 4.4.2)
callr	3.7.6	CRAN (R 4.4.2)
car	3.1-3	CRAN (R 4.4.2)
carData	3.0-5	CRAN (R 4.4.2)
cellranger	1.1.0	CRAN (R 4.4.2)
cli	3.6.3	CRAN (R 4.4.2)
coda	0.19-4.1	CRAN (R 4.4.2)
colorspace	2.1-1	CRAN (R 4.4.2)
corpcor	1.6.10	CRAN (R 4.4.0)
crayon	1.5.3	CRAN (R 4.4.2)
cubature	2.1.2	CRAN (R 4.4.3)
curl	6.2.2	CRAN (R 4.4.3)
data.table	1.16.4	CRAN (R 4.4.2)
DBI	1.2.3	CRAN (R 4.4.2)
dbplyr	2.5.0	CRAN (R 4.4.2)
desc	1.4.3	CRAN (R 4.4.2)
devtools	2.4.5	CRAN (R 4.4.2)

digest	0.6.37	CRAN (R 4.4.2)
dplyr	1.1.4	CRAN (R 4.4.2)
ellipsis	0.3.2	CRAN (R 4.4.2)
evaluate	1.0.3	CRAN (R 4.4.2)
extrafont	NA	NA
extrafontdb	NA	NA
fansi	1.0.6	CRAN (R 4.4.2)
faraway	1.0.9	CRAN (R 4.4.3)
fastmap	1.2.0	CRAN (R 4.4.2)
forcats	1.0.0	CRAN (R 4.4.2)
foreign	0.8-90	CRAN (R 4.4.3)
fs	1.6.6	CRAN (R 4.4.3)
generics	0.1.3	CRAN (R 4.4.2)
ggplot2	3.5.2	CRAN (R 4.4.3)
glue	1.8.0	CRAN (R 4.4.2)
gtable	0.3.6	CRAN (R 4.4.2)
haven	2.5.4	CRAN (R 4.4.2)
Hmisc	5.2-3	CRAN (R 4.4.3)
hms	1.1.3	CRAN (R 4.4.2)
htmltools	0.5.8.1	CRAN (R 4.4.2)
htmlwidgets	1.6.4	CRAN (R 4.4.2)
httr	1.4.7	CRAN (R 4.4.2)
investr	1.4.2	CRAN (R 4.4.3)
jpeg	0.1-11	CRAN (R 4.4.3)
jquerylib	0.1.4	CRAN (R 4.4.2)
jsonlite	2.0.0	CRAN (R 4.4.3)
kableExtra	1.4.0	CRAN (R 4.4.3)
knitr	1.50	CRAN (R 4.4.3)
lattice	0.22-6	CRAN (R 4.4.3)
latticeExtra	0.6-30	CRAN (R 4.4.2)
lifecycle	1.0.4	CRAN (R 4.4.2)
lme4	1.1-37	CRAN (R 4.4.3)
lmerTest	3.1-3	CRAN (R 4.4.2)
lsr	0.5.2	CRAN (R 4.4.3)
ltm	1.2-0	CRAN (R 4.4.3)
lubridate	1.9.4	CRAN (R 4.4.2)
magrittr	2.0.3	CRAN (R 4.4.2)
MASS	7.3-65	CRAN (R 4.4.3)
matlib	1.0.0	CRAN (R 4.4.3)
Matrix	1.7-2	CRAN (R 4.4.3)
MCMCglmm	2.36	CRAN (R 4.4.2)
memoise	2.0.1	CRAN (R 4.4.2)
mgcv	1.9-3	CRAN (R 4.4.3)

(continued)

package	version	source
minqa	1.2.8	CRAN (R 4.4.2)
modelr	0.1.11	CRAN (R 4.4.2)
munsell	0.5.1	CRAN (R 4.4.2)
nlme	3.1-167	CRAN (R 4.4.3)
nloptr	2.2.1	CRAN (R 4.4.3)
nlstools	2.1-0	CRAN (R 4.4.2)
nnet	7.3-20	CRAN (R 4.4.3)
numDeriv	2016.8-1.1	CRAN (R 4.4.0)
openxlsx	4.2.8	CRAN (R 4.4.3)
pbkrtest	0.5.4	CRAN (R 4.4.3)
pillar	1.10.2	CRAN (R 4.4.3)
pkgbuild	1.4.7	CRAN (R 4.4.3)
pkgconfig	2.0.3	CRAN (R 4.4.2)
pkgload	1.4.0	CRAN (R 4.4.2)
png	0.1-8	CRAN (R 4.4.0)
ppsr	0.0.5	CRAN (R 4.4.3)
prettyunits	1.2.0	CRAN (R 4.4.2)
processx	3.8.6	CRAN (R 4.4.3)
ps	1.9.1	CRAN (R 4.4.3)
pscl	1.5.9	CRAN (R 4.4.2)
purrr	1.0.2	CRAN (R 4.4.2)
R6	2.6.1	CRAN (R 4.4.3)
RColorBrewer	1.1-3	CRAN (R 4.4.0)
Rcpp	1.0.14	CRAN (R 4.4.2)
readr	2.1.5	CRAN (R 4.4.2)
readxl	1.4.5	CRAN (R 4.4.3)
remotes	2.5.0	CRAN (R 4.4.2)
reprex	2.1.1	CRAN (R 4.4.2)
rgl	1.3.18	CRAN (R 4.4.3)
rio	1.2.3	CRAN (R 4.4.2)
rlang	1.1.5	CRAN (R 4.4.2)
RLRsim	3.1-8	CRAN (R 4.4.3)
rmarkdown	2.29	CRAN (R 4.4.2)
rprojroot	2.0.4	CRAN (R 4.4.2)
rstudioapi	0.17.1	CRAN (R 4.4.2)
Rttf2pt1	NA	NA
rvest	1.0.4	CRAN (R 4.4.2)
sass	0.4.10	CRAN (R 4.4.3)
scales	1.4.0	CRAN (R 4.4.3)
sessioninfo	1.2.3	CRAN (R 4.4.3)
stringi	1.8.7	CRAN (R 4.4.3)
stringr	1.5.1	CRAN (R 4.4.2)

svglite	2.2.0	CRAN (R 4.4.3)
systemfonts	1.2.3	CRAN (R 4.4.3)
tensorA	0.36.2.1	CRAN (R 4.4.0)
testthat	3.2.3	CRAN (R 4.4.2)
tibble	3.2.1	CRAN (R 4.4.2)
tidyr	1.3.1	CRAN (R 4.4.2)
tidyselect	1.2.1	CRAN (R 4.4.2)
tidyverse	2.0.0	CRAN (R 4.4.2)
tzdb	0.5.0	CRAN (R 4.4.3)
usethis	3.1.0	CRAN (R 4.4.2)
utf8	1.2.5	CRAN (R 4.4.3)
vctrs	0.6.5	CRAN (R 4.4.2)
viridisLite	0.4.2	CRAN (R 4.4.2)
webshot	0.5.5	CRAN (R 4.4.3)
withr	3.0.2	CRAN (R 4.4.2)
xfun	0.51	CRAN (R 4.4.3)
xml2	1.3.8	CRAN (R 4.4.3)
xtable	1.8-4	CRAN (R 4.4.2)
yaml	2.3.10	CRAN (R 4.4.2)
zip	2.3.2	CRAN (R 4.4.3)

```
#> - Session info ---------------------------------------------------------------
#>  setting  value
#>  version  R version 4.4.3 (2025-02-28 ucrt)
#>  os       Windows 11 x64 (build 26100)
#>  system   x86_64, mingw32
#>  ui       RTerm
#>  language (EN)
#>  collate  English_United States.utf8
#>  ctype    English_United States.utf8
#>  tz       America/Los_Angeles
#>  date     2025-05-10
#>  pandoc   3.1.1 @ C:/Program Files/RStudio/resources/app/bin/quarto/bin/tools/ (via
          rmarkdown)
#>  quarto   NA @ C:\\PROGRA~1\\RStudio\\resources\\app\\bin\\quarto\\bin\\quarto.exe
#>
#> - Packages -------------------------------------------------------------------
#>  package      * version date (UTC) lib source
#>  bookdown       0.43    2025-04-15 [1] CRAN (R 4.4.3)
#>  cachem         1.1.0   2024-05-16 [1] CRAN (R 4.4.2)
#>  cli            3.6.3   2024-06-21 [1] CRAN (R 4.4.2)
#>  codetools      0.2-20  2024-03-31 [1] CRAN (R 4.4.3)
#>  desc           1.4.3   2023-12-10 [1] CRAN (R 4.4.2)
#>  devtools       2.4.5   2022-10-11 [1] CRAN (R 4.4.2)
#>  dichromat      2.0-0.1 2022-05-02 [1] CRAN (R 4.4.0)
#>  digest         0.6.37  2024-08-19 [1] CRAN (R 4.4.2)
#>  dplyr        * 1.1.4   2023-11-17 [1] CRAN (R 4.4.2)
#>  ellipsis       0.3.2   2021-04-29 [1] CRAN (R 4.4.2)
#>  evaluate       1.0.3   2025-01-10 [1] CRAN (R 4.4.2)
#>  farver         2.1.2   2024-05-13 [1] CRAN (R 4.4.2)
#>  fastmap        1.2.0   2024-05-15 [1] CRAN (R 4.4.2)
#>  forcats      * 1.0.0   2023-01-29 [1] CRAN (R 4.4.2)
#>  fs             1.6.6   2025-04-12 [1] CRAN (R 4.4.3)
#>  generics       0.1.3   2022-07-05 [1] CRAN (R 4.4.2)
```

```
#>  ggplot2      * 3.5.2   2025-04-09 [1] CRAN (R 4.4.3)
#>  glue           1.8.0   2024-09-30 [1] CRAN (R 4.4.2)
#>  gtable         0.3.6   2024-10-25 [1] CRAN (R 4.4.2)
#>  hms            1.1.3   2023-03-21 [1] CRAN (R 4.4.2)
#>  htmltools      0.5.8.1 2024-04-04 [1] CRAN (R 4.4.2)
#>  htmlwidgets    1.6.4   2023-12-06 [1] CRAN (R 4.4.2)
#>  httpuv         1.6.16  2025-04-16 [1] CRAN (R 4.4.3)
#>  knitr          1.50    2025-03-16 [1] CRAN (R 4.4.3)
#>  later          1.4.2   2025-04-08 [1] CRAN (R 4.4.3)
#>  lifecycle      1.0.4   2023-11-07 [1] CRAN (R 4.4.2)
#>  lubridate    * 1.9.4   2024-12-08 [1] CRAN (R 4.4.2)
#>  magrittr       2.0.3   2022-03-30 [1] CRAN (R 4.4.2)
#>  memoise        2.0.1   2021-11-26 [1] CRAN (R 4.4.2)
#>  mime           0.13    2025-03-17 [1] CRAN (R 4.4.3)
#>  miniUI         0.1.2   2025-04-17 [1] CRAN (R 4.4.3)
#>  pillar         1.10.2  2025-04-05 [1] CRAN (R 4.4.3)
#>  pkgbuild       1.4.7   2025-03-24 [1] CRAN (R 4.4.3)
#>  pkgconfig      2.0.3   2019-09-22 [1] CRAN (R 4.4.2)
#>  pkgload        1.4.0   2024-06-28 [1] CRAN (R 4.4.2)
#>  profvis        0.4.0   2024-09-20 [1] CRAN (R 4.4.2)
#>  promises       1.3.2   2024-11-28 [1] CRAN (R 4.4.2)
#>  purrr        * 1.0.2   2023-08-10 [1] CRAN (R 4.4.2)
#>  R6             2.6.1   2025-02-15 [1] CRAN (R 4.4.3)
#>  RColorBrewer   1.1-3   2022-04-03 [1] CRAN (R 4.4.0)
#>  Rcpp           1.0.14  2025-01-12 [1] CRAN (R 4.4.2)
#>  readr        * 2.1.5   2024-01-10 [1] CRAN (R 4.4.2)
#>  remotes        2.5.0   2024-03-17 [1] CRAN (R 4.4.2)
#>  rlang          1.1.5   2025-01-17 [1] CRAN (R 4.4.2)
#>  rmarkdown      2.29    2024-11-04 [1] CRAN (R 4.4.2)
#>  rstudioapi     0.17.1  2024-10-22 [1] CRAN (R 4.4.2)
#>  scales       * 1.4.0   2025-04-24 [1] CRAN (R 4.4.3)
#>  sessioninfo    1.2.3   2025-02-05 [1] CRAN (R 4.4.3)
#>  shiny          1.10.0  2024-12-14 [1] CRAN (R 4.4.2)
#>  stringi        1.8.7   2025-03-27 [1] CRAN (R 4.4.3)
#>  stringr      * 1.5.1   2023-11-14 [1] CRAN (R 4.4.2)
#>  tibble       * 3.2.1   2023-03-20 [1] CRAN (R 4.4.2)
#>  tidyr        * 1.3.1   2024-01-24 [1] CRAN (R 4.4.2)
#>  tidyselect     1.2.1   2024-03-11 [1] CRAN (R 4.4.2)
#>  tidyverse    * 2.0.0   2023-02-22 [1] CRAN (R 4.4.2)
#>  timechange     0.3.0   2024-01-18 [1] CRAN (R 4.4.2)
#>  tinytex        0.57    2025-04-15 [1] CRAN (R 4.4.3)
#>  tzdb           0.5.0   2025-03-15 [1] CRAN (R 4.4.3)
#>  urlchecker     1.0.1   2021-11-30 [1] CRAN (R 4.4.2)
#>  usethis        3.1.0   2024-11-26 [1] CRAN (R 4.4.2)
#>  vctrs          0.6.5   2023-12-01 [1] CRAN (R 4.4.2)
#>  withr          3.0.2   2024-10-28 [1] CRAN (R 4.4.2)
#>  xfun           0.51    2025-02-19 [1] CRAN (R 4.4.3)
#>  xtable         1.8-4   2019-04-21 [1] CRAN (R 4.4.2)
#>  yaml           2.3.10  2024-07-26 [1] CRAN (R 4.4.2)
#>
#>  [1] C:/Program Files/R/R-4.4.3/library
#>  * -- Packages attached to the search path.
#>
#> ------------------------------------------------------------------------------
```

Chapter 2
Causal Inference

Throughout our journey into statistical concepts, we've uncovered patterns, relationships, and trends in data. But now, we arrive at one of the most profound questions in all of research and decision-making: **What truly causes what?**

We've all heard the phrase.

Correlation is not causation.

Just because two things move together doesn't mean one is pulling the strings of the other. Ice cream sales and drowning incidents both rise in the summer, but ice cream isn't to blame.

But what exactly *is* causation? Let's explore.

One of the most insightful books on this topic is *The Book of Why* by Judea Pearl (Pearl & Mackenzie, 2018), which explains the nuances of causal reasoning beautifully. We'll briefly visit a concise summary of key ideas from Pearl's work, supplemented with insights from econometrics and statistics.

Understanding causal relationships is essential in research, particularly in fields like economics, finance, marketing, and medicine. While statistical methods have traditionally focused on associational reasoning, causal inference allows us to answer **what-if questions** and make decisions based on interventions.

However, one must be aware of the limitations of statistical methods. As discussed throughout this book, relying solely on data without incorporating domain expertise can lead to misleading conclusions. To establish causality, we often need expert judgment, prior research, and rigorous experimental design.

You may have come across amusing examples of **spurious correlations**—such as the famous Tyler Vigen collection, which shows absurd relationships (e.g., "the number of Nicholas Cage movies correlates with drowning accidents"). These highlight the danger of mistaking correlation for causation.

Historically, one of the earliest attempts to infer causation using **regression analysis** was by Yule (1899), who investigated the effect of relief policies on poverty. Unfortunately, his analysis suggested that relief policies increased poverty—a misleading conclusion due to unaccounted confounders.

M. Nguyen, *Experimental Design*, https://doi.org/10.1007/978-3-032-01839-7_2

For a long time, statistics was largely a **causality-free discipline**. The field only began addressing causation in the 1920s, when Sewall Wright introduced **path analysis**, a graphical approach to representing causal relationships (Wright, 1920). However, it wasn't until Judea Pearl's **Causal Revolution** (1990s) that we gained a formal calculus for causation.

Pearl's framework introduced two key innovations:

1. **Causal Diagrams** (Directed Acyclic Graphs)—A graphical representation of cause-and-effect relationships
2. **A Symbolic Language**: The Do-Operator ($do(X)$)—A mathematical notation for interventions

Traditional statistics deals with **conditional probabilities**:

$$P(Y|X)$$

This formula tells us the probability of event Y occurring given that event X has occurred. In the context of observed data, $P(Y \mid X)$ reflects the association between X and Y, showing how likely Y is when X happens.

However, causal inference requires a different concept:

$$P(Y|do(X))$$

which describes what happens when we **actively intervene** and set X. The crucial distinction is:

$$P(Y|X) \neq P(Y|do(X))$$

in general, because **passively observing** X is not the same as **actively manipulating** it.

To make causal claims, we need to answer **counterfactual questions**:

What would have happened if we had NOT done X?

This concept is essential in fields like policy evaluation, medicine, and business decision-making.

Pearl and Mackenzie (2018) outline **three levels of cognitive ability** required for causal learning:

1. **Seeing**—Observing associations in data
2. **Doing**—Understanding interventions and predicting their outcomes
3. **Imagining**—Reasoning about counterfactuals

This is also known as Pearl's *Ladder of Causation* that is used to describe three hierarchical levels of causal reasoning.

Each level requires more cognitive ability and data. Classical statistics operates at Level 1 (association), while causal inference enables us to reach Levels 2 and 3.

2.1 The Formal Notation of Causality

A common mistake is defining causation using probability:

$$X \text{ causes } Y \text{ if } P(Y|X) > P(Y).$$

Seeing X (first level) **doesn't mean** the probability of Y increases.
It could be either that

1. X causes Y, or
2. Z affects both X and Y. We might be able to use **control variables**—$P(Y|X, Z = z) > P(Y|Z = z)$. But then the question becomes
   (a) How to choose Z?
   (b) Did you choose enough Z?
   (c) Did you choose the right Z?

Hence, the previous statement is incorrect. The correct causal statement is:

$$P(Y|do(X)) > P(Y).$$

With **causal diagrams** and **do-calculus**, we can formally express interventions and answer questions at the second level (Intervention).

Pearl (2019) also introduces the Pearl's *Structural Causal Model (SCM)* framework for causal inference (Pearl, 2019):

1. **Encoding Causal Assumptions**—Using **causal graphs** for transparency and testability
2. **Do-Calculus**—Controlling for confounding using the **backdoor criterion**
3. **Algorithmization of Counterfactuals**—Modeling "what if?" scenarios
4. **Mediation Analysis**—Understanding direct vs. indirect effects
5. **External Validity and Adaptability**—Addressing selection bias and domain adaptation
6. **Handling Missing Data**—Using causal methods to infer missing information
7. **Causal Discovery**—Learning causal relationships from data using:
   - **d-separation**
   - **Functional decomposition** (Hoyer et al., 2008)
   - **Spontaneous local changes** (Pearl, 2014)

To explore causal inference in R, check out the CRAN Task View for Causal Inference:

For further reading:

- *The Book of Why*—Judea Pearl (Pearl & Mackenzie, 2018)
- *Causal Inference in Statistics: A Primer*—Pearl, Glymour, Jewell
- *Causality: Models, Reasoning, and Inference*—Judea Pearl

2.2 Simpson's Paradox

Simpson's Paradox is one of the most striking examples of **why causality matters** and why simple statistical associations can be misleading.

At its core, Simpson's Paradox occurs when:

> A trend observed in an overall population **reverses** when the population is divided into subgroups.

This means that **statistical associations in raw data can be misleading** if important confounding variables are ignored.

Understanding Simpson's Paradox is critical in causal inference because:

1. It highlights the danger of naive data analysis—Just looking at overall trends can lead to incorrect conclusions.
2. It emphasizes the importance of confounding variables—We must control for relevant factors before making causal claims.
3. It demonstrates why causal reasoning is necessary—Simply relying on statistical associations ($P(Y|X)$) without considering structural relationships can lead to paradoxical results.

2.2.1 Comparison Between Simpson's Paradox and Omitted Variable Bias

Simpson's Paradox occurs when a trend in an overall dataset **reverses** when broken into subgroups. This happens due to **data aggregation issues**, where differences in subgroup sizes distort the overall trend.

While this often resembles **omitted variable bias (OVB)**—where missing confounders lead to misleading conclusions—Simpson's Paradox is not just a causal inference problem. It is a **mathematical phenomenon** that can arise purely from **improper weighting of data**, even in descriptive statistics.

Similarities Between Simpson's Paradox and OVB

1. Both involve a missing variable:
    - In Simpson's Paradox, a key confounding variable (e.g., customer segment) is hidden in the aggregate data, leading to misleading conclusions.
    - In OVB, a relevant variable (e.g., seasonality) is missing from the regression model, causing bias.
2. Both distort causal conclusions:
    - OVB effect estimates by failing to control for confounding.
    - Simpson's Paradox flips statistical relationships when controlling for a confounder.

Differences Between Simpson's Paradox and OVB

1. Not all OVB cases show Simpson's Paradox:
   - OVB generally causes bias, but it doesn't always create a reversal of trends.
   - Example: If seasonality increases both ad spend and sales, omitting it inflates the ad spend → sales relationship but does not necessarily reverse it.
2. Simpson's Paradox can occur even without causal inference:
   - Simpson's Paradox is a mathematical/statistical phenomenon that can arise even in purely observational data, not just causal inference.
   - It results from data weighting issues, even if causality is not considered.
3. OVB is a model specification issue; Simpson's Paradox is a data aggregation issue:
   - OVB occurs in regression models when we fail to include relevant predictors.
   - Simpson's Paradox arises from incorrect data aggregation when groups are not properly analyzed separately.

The Right Way to Think About It

- Simpson's Paradox is often *caused* by omitted variable bias, but they are not the same thing.
- OVB is a problem in causal inference models; Simpson's Paradox is a problem in raw data interpretation.

How to Fix These Issues?

- For OVB: Use causal diagrams, add control variables, and use regression adjustments.
- For Simpson's Paradox: Always analyze subgroup-level trends before making conclusions based on aggregate data.
- Bottom line: Simpson's Paradox is often *caused* by omitted variable bias, but it is not just OVB—it is a fundamental issue of misleading data aggregation.

2.2.2 *Illustrating Simpson's Paradox: Marketing Campaign Success Rates*

Let's explore this paradox using a practical business example.

Scenario: Marketing Campaign Performance

Imagine a company running two marketing campaigns, Campaign A and Campaign B, to attract new customers. We analyze which campaign has a higher conversion rate.

Step 1: Creating the Data
We will simulate conversion rates for two different customer segments:

1. **High-Value** customers (who typically convert at a higher rate)
2. **Low-Value** customers (who convert at a lower rate)

```r
# Load necessary libraries
library(dplyr)

# Create a dataset where:
#   - B is better than A in each individual segment.
#   - A turns out better when we look at the overall
↪   (aggregated) data.

marketing_data <- data.frame(
  Campaign = c("A", "A", "B", "B"),
  Segment  = c("High-Value", "Low-Value", "High-Value",
  ↪   "Low-Value"),
  Visitors = c(500, 2000, 300, 3000),     # total visitors in
  ↪   each segment
  Conversions = c(290, 170, 180, 270)    # successful
  ↪   conversions
)

# Compute segment-level conversion rate
marketing_data <- marketing_data %>%
  mutate(Conversion_Rate = Conversions / Visitors)

# Print the data
print(marketing_data)
#>   Campaign    Segment Visitors Conversions Conversion_Rate
#> 1        A High-Value      500         290           0.580
#> 2        A  Low-Value     2000         170           0.085
#> 3        B High-Value      300         180           0.600
#> 4        B  Low-Value     3000         270           0.090
```

Interpreting This Data

- **Campaign B** in the High-Value segment: $\frac{180}{300} = 60\%$
- **Campaign A** in the High-Value segment: $\frac{290}{500} = 58\%$
  => B is better in the High-Value segment (60% vs 58%).
- **Campaign B** in the Low-Value segment: $\frac{270}{3000} = 9\%$
- **Campaign A** in the Low-Value segment: $\frac{170}{2000} = 8.5\%$
  => B is better in the Low-Value segment (9% vs 8.5%).

Thus, **B** outperforms **A** in each **individual** segment.

Step 2: Aggregating Data (Ignoring Customer Segments)
Now, let's calculate the overall conversion rate for each campaign without considering customer segments.

```
# Compute overall conversion rates for each campaign
overall_rates <- marketing_data %>%
  group_by(Campaign) %>%
  summarise(
    Total_Visitors          = sum(Visitors),
    Total_Conversions       = sum(Conversions),
    Overall_Conversion_Rate = Total_Conversions /
    ↪  Total_Visitors
  )

# Print overall conversion rates
print(overall_rates)
#> # A tibble: 2 x 4
#>   Campaign Total_Visitors Total_Conversions
↪  Overall_Conversion_Rate
#>   <chr>             <dbl>             <dbl>
↪  <dbl>
#> 1 A                  2500               460
↪  0.184
#> 2 B                  3300               450
↪  0.136
```

Step 3: Observing Simpson's Paradox
Let's determine which campaign appears to have a higher conversion rate.

```
# Identify the campaign with the higher overall conversion
↪  rate
best_campaign_overall <- overall_rates %>%
    filter(Overall_Conversion_Rate ==
    ↪  max(Overall_Conversion_Rate)) %>%
    select(Campaign, Overall_Conversion_Rate)

print(best_campaign_overall)
#> # A tibble: 1 x 2
#>   Campaign Overall_Conversion_Rate
#>   <chr>                      <dbl>
#> 1 A                          0.184
```

Even though **Campaign B** is **better** in each **segment**, you should see that **Campaign A** has a **higher** aggregated (overall) conversion rate!

Step 4: Conversion Rates Within Customer Segments
We now analyze the conversion rates separately for high-value and low-value customers.

```
# Compute conversion rates by customer segment
by_segment <- marketing_data %>%
  select(Campaign, Segment, Conversion_Rate) %>%
  arrange(Segment)

print(by_segment)
#>   Campaign    Segment Conversion_Rate
#> 1        A High-Value           0.580
#> 2        B High-Value           0.600
#> 3        A  Low-Value           0.085
#> 4        B  Low-Value           0.090
```

- In **High-Value**, B > A.
- In **Low-Value**, B > A.

Yet, overall, A > B.
This **reversal** is the hallmark of **Simpson's Paradox**.

Step 5: Visualizing the Paradox
To make this clearer, let's visualize the results.

```
library(ggplot2)

# Plot conversion rates by campaign and segment
ggplot(marketing_data,
       aes(x = Segment,
           y = Conversion_Rate,
           fill = Campaign)) +
  geom_bar(stat = "identity", position = "dodge") +
  labs(
    title = "Simpson's Paradox in Marketing",
    x     = "Customer Segment",
    y     = "Conversion Rate"
  ) +
  theme_minimal()
```

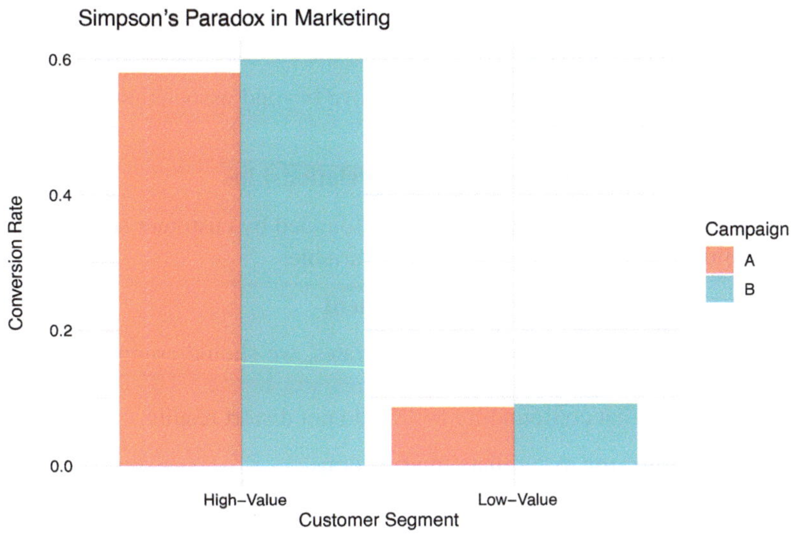

Fig. 2.1 Simpson Paradox in marketing

This bar chart reveals that for **both** segments, B's bar is taller (i.e., B's conversion rate is higher). If you only examined segment-level data, you would conclude that B is the superior campaign (Fig. 2.1).

However, if you aggregate the data (ignore segments), you get the opposite conclusion—that **A** is better overall.

2.2.3 Why Does This Happen?

This paradox arises because of a **confounding variable**—in this case, the **distribution of visitors** across segments.

- **Campaign A** has **more** of its traffic in the High-Value segment (where conversions are generally high).
- **Campaign B** has **many** of its visitors in the Low-Value segment.

Because the **volume** of Low-Value visitors in B is extremely large (3000 vs. 2000 for A), it weighs down B's **overall** average more, allowing A's overall rate to exceed B's.

2.2.4 How Does Causal Inference Solve This?

To avoid Simpson's Paradox, we need to move beyond association and use causal analysis:

1. **Use causal diagrams (DAGs) to model relationships**
   - The marketing campaign choice is confounded by customer segment.
   - We must control for the confounding variable.
2. **Use stratification or regression adjustment**
   - Instead of comparing raw conversion rates, we should compare rates within each customer segment.
   - This ensures that confounding factors do not distort results.
3. **Use the do-operator** to simulate interventions
   - Instead of asking $P(\text{Conversion} \mid \text{Campaign})$, ask: $P(\text{Conversion} \mid do(\text{Campaign}))$
   - This estimates what would happen if we randomly assigned campaigns (removing confounding bias).

2.2.5 Correcting Simpson's Paradox with Regression Adjustment

Let's adjust for the confounding variable using **logistic regression**.

```
# Logistic regression adjusting for the Segment
model <- glm(
  cbind(Conversions, Visitors - Conversions) ~ Campaign +
  ↪  Segment,
  family = binomial(),
  data   = marketing_data
)

summary(model)
#>
#> Call:
#> glm(formula = cbind(Conversions, Visitors - Conversions) ~
↪  Campaign +
#>     Segment, family = binomial(), data = marketing_data)
#>
#> Coefficients:
```

```
#>                   Estimate Std. Error z value Pr(>|z|)
#> (Intercept)        0.32783    0.07839   4.182 2.89e-05 ***
#> CampaignB          0.06910    0.08439   0.819    0.413
#> SegmentLow-Value -2.70806    0.08982 -30.151  < 2e-16 ***
#> ---
#> Signif. codes:  0 '***' 0.001 '**' 0.01 '*' 0.05 '.' 0.1 ' ' 1
#>
#> (Dispersion parameter for binomial family taken to be 1)
#>
#>     Null deviance: 977.473003  on 3  degrees of freedom
#> Residual deviance:   0.012337  on 1  degrees of freedom
#> AIC: 32.998
#>
#> Number of Fisher Scoring iterations: 3
```

This model includes both **Campaign** and **Segment** as predictors, giving a clearer picture of the *true effect* of each campaign on conversion, **after** controlling for differences in segment composition.

2.2.6 *Key Takeaways*

1. **Simpson's Paradox demonstrates why causal inference is essential.**
   - Aggregated statistics can be misleading due to hidden confounding.
   - Breaking data into subgroups can reverse conclusions.
2. **Causal reasoning helps identify and correct paradoxes.**
   - Using causal graphs, do-calculus, and adjustment techniques, we can find the true causal effect.
3. **Naïve data analysis can lead to bad business decisions.**
   - If a company allocated more budget to Campaign B based on overall conversion rates, it might be investing in the wrong strategy!

2.3 Experimental Versus Quasi-Experimental Designs

To determine whether a particular intervention or treatment causes an observed outcome, it requires more than observing associations—it demands a framework for **causal inference**.

Table 2.1 Experimental design versus quasi-experimental design

Feature	Experimental design	Quasi-experimental design
Assignment to treatment	Randomized	Nonrandomized (observational)
Control over confounding	High	Limited (requires statistical control)
Causal inference validity	Strong (if properly implemented)	Weaker (depends on assumptions)
Feasibility in field studies	Often difficult or costly	More flexible and practical
Examples	A/B testing, clinical trials	Difference-in-differences, matching
Principal investigator	Conducted by an experimentalist	Conducted by an observationalist
Type of data	Uses experimental data	Uses observational data
Randomness helps	Random assignment reduces treatment imbalance	Random sampling reduces sample selection error

To address this, researchers rely on two broad classes of research designs: **experimental** and **quasi-experimental**. Both aim to estimate causal effects, but they differ significantly in their level of control over the assignment mechanism and the assumptions required for valid inference.

- Experimental Design, particularly randomized controlled trials (RCTs), is considered the gold standard for causal inference. By randomly assigning units (e.g., customers, users, regions) to treatment or control groups, these designs eliminate confounding and allow for straightforward interpretation of treatment effects. However, in many business settings, true randomization is costly, impractical, or ethically constrained.
- Quasi-Experimental designs provide an alternative when random assignment is not feasible. These designs rely on observational data and statistical techniques to approximate the conditions of an experiment. While more flexible in application, they typically require stronger assumptions and careful methodological implementation to yield credible causal insights.

Table 2.1 summarizes the key differences between these two approaches, highlighting their respective strengths, limitations, and use cases in applied research.

2.3.1 Criticisms of Quasi-Experimental Designs

Quasi-experimental methods do not always approximate experimental results accurately. For instance, LaLonde (1986) demonstrates that commonly used methods such as:

- [Matching Methods]

- [Difference-in-differences]
- [Tobit-2] (Heckman-type models)

often fail to replicate experimental estimates reliably. This finding casts serious doubt on the credibility of observational studies for estimating causal effects, igniting an ongoing debate in econometrics and statistics about the reliability of nonexperimental evaluations.

LaLonde's critical assessment served as a catalyst for significant methodological and practical advancements in causal inference. In the decades since this publication, the field has evolved considerably, introducing both theoretical innovations and empirical practices aimed at addressing the limitations that were exposed (Imbens & Xu, 2024). Among these advances are:

- **Emphasis on estimators based on unconfoundedness (selection on observables):** Modern causal inference frameworks frequently adopt the *unconfoundedness* or *conditional independence* assumption. Under this premise, treatment assignment is assumed to be independent of potential outcomes, conditional on observed covariates. This theoretical foundation underpins many widely used estimation techniques, such as matching methods, inverse probability weighting, and regression adjustment.
- **Focus on covariate overlap (common support):** Researchers now recognize the critical importance of *overlap*, also referred to as *common support*, in the distributions of covariates across treatment and control groups. Without sufficient overlap, comparisons between treated and untreated units rely on extrapolation, which weakens causal claims. Modern methods explicitly assess and often impose restrictions to ensure overlap before proceeding with estimation.
- **Introduction of propensity score-based methods and doubly robust estimators:** The introduction of *propensity score* methods (Rosenbaum & Rubin, 1983) was a breakthrough, offering a way to reduce the dimensionality of the covariate space while balancing observed characteristics across groups. More recently, *doubly robust* estimators have emerged, combining propensity score weighting with outcome regression. These estimators provide consistent treatment effect estimates if either the propensity score model or the outcome model is correctly specified, offering greater robustness in practice.
- **Greater emphasis on validation exercises to bolster credibility:** Modern studies increasingly incorporate validation techniques to evaluate the credibility of their findings. *Placebo tests*, *falsification exercises*, and *sensitivity analyses* are commonly employed to assess whether estimated effects may be driven by unobserved confounding or model misspecification. Such practices go beyond traditional goodness-of-fit statistics, directly interrogating the assumptions underlying causal inference.
- **Methods for estimating and exploiting treatment effect heterogeneity:** Beyond estimating average treatment effects, contemporary research frequently explores *heterogeneous treatment effects*. These methods identify subgroups that may experience different causal impacts, which is of particular relevance in fields like personalized marketing, targeted interventions, and policy design.

To illustrate the practical lessons from these methodological advances, Imbens and Xu (2024) reexamine two canonical datasets:

1. **LaLonde's National Supported Work Demonstration data**
2. **The Imbens-Rubin-Sacerdote draft lottery data**

Applying modern causal inference methods to these datasets demonstrates that, when sufficient covariate overlap exists, robust estimates of the adjusted differences between treatment and control groups can be achieved. However, it is critical to underscore that robustness in estimation does not equate to validity. Without direct validation exercises, such as placebo tests, even well-behaved estimates may be misleading.

Imbens and Xu (2024) also highlight several key lessons for practitioners working with nonexperimental data to estimate causal effects:

- **Careful examination of the assignment process is essential.**
  Understanding the mechanisms by which units are assigned to treatment or control conditions informs the plausibility of the unconfoundedness assumption.
- **Inspection of covariate overlap is nonnegotiable.**
  Without sufficient overlap, causal effect estimation may rely heavily on model extrapolation, undermining credibility.
- **Validation exercises are indispensable.**
  Placebo tests and falsification strategies help ensure that estimated treatment effects are not artifacts of modeling choices or unobserved confounding.

While methodological advances have substantially improved the tools available for causal inference with observational data, their effective application requires rigorous attention to the underlying assumptions and diligent validation to support credible causal claims.

2.4 Hierarchical Ordering of Causal Tools

Causal inference tools can be categorized based on their methodological rigor, with randomized controlled trials (RCTs) considered the gold standard.

1. Experimental Design: Randomized Control Trials (Gold standard)
2. Quasi-experimental
    (a) Regression Discontinuity
    (b) Synthetic Difference-in-Differences
    (c) Difference-In-Differences
    (d) Synthetic Control
    (e) Event Studies
    (f) Fixed Effects Estimator
    (g) Endogenous Treatment]: mostly [Instrumental Variables
    (h) Matching Methods

(i) Interrupted Time Series
(j) Endogenous Sample Selection

2.5 Types of Validity in Research

Validity in research includes:

1. Measurement validity (e.g., construct, content, criterion, face validity)
2. Internal validity
3. External validity
4. Ecological validity
5. Statistical conclusion validity

By examining these, you can ensure that your study's measurements are accurate, your findings are reliably causal, and your conclusions generalize to broader contexts.

2.5.1 Measurement Validity

Measurement validity pertains to whether the instrument or method you use truly measures what it's intended to measure. Within this umbrella, there are several subtypes:

2.5.1.1 Face Validity

- **Definition**: The extent to which a measurement or test *appears* to measure what it is supposed to measure, at face value (i.e., does it "look" right to experts or users?).
- **Importance**: While often considered a less rigorous form of validity, it's useful for ensuring the test or instrument is intuitively acceptable to stakeholders, participants, or experts in the field.
- **Example**: A questionnaire measuring "anxiety" that has questions about nervousness, worries, and stress has good face validity because it obviously seems to address anxiety.

2.5.1.2 Content Validity

- **Definition**: The extent to which a test or measurement covers *all* relevant facets of the construct it aims to measure.

- **Importance**: Especially critical in fields like education or psychological testing, where you want to ensure the entire domain of a subject/construct is properly sampled.
- **Example**: A math test that includes questions on algebra, geometry, and calculus might have high content validity for a comprehensive math skill assessment. If it only tested algebra, the content validity would be low.

2.5.2 *Construct Validity*

- **Definition**: The degree to which a test or measurement tool accurately represents the theoretical construct it intends to measure (e.g., intelligence, motivation, self-esteem).
- **Types of Evidence**:
  - **Convergent Validity**: Demonstrated when measures that are supposed to be related (theoretically) are observed to correlate.
  - **Discriminant (Divergent) Validity**: Demonstrated when measures that are supposed to be unrelated theoretically do not correlate.
- **Example**: A new questionnaire on "job satisfaction" should correlate with other established job satisfaction questionnaires (convergent validity) but should not correlate strongly with unrelated constructs like "physical health" (discriminant validity).

2.5.3 *Criterion Validity*

- **Definition**: The extent to which the measurement predicts or correlates with an outcome criterion. In other words, do scores on the measure relate to an external standard or "criterion"?
- **Types**:
  - **Predictive Validity**: The measure predicts a future outcome (e.g., an entrance exam predicting college success).
  - **Concurrent Validity**: The measure correlates with an existing, accepted measure taken at the same time (e.g., a new depression scale compared with a gold-standard clinical interview).
- **Example**: A new test of driving skills has high criterion validity if people who score highly perform better on actual road tests (predictive validity).

2.5.4 *Internal Validity*

Internal validity refers to the extent to which a study can establish a *cause-and-effect* relationship. High internal validity means you can be confident that the observed effects are due to the treatment or intervention itself and not due to confounding factors or alternative explanations. This is the validity that economists and applied scientists largely care about.

2.5.4.1 Major Threats to Internal Validity

1. **Selection Bias**: Systematic differences between groups that exist before the treatment is applied.
2. **History Effects**: External events occurring during the study can affect outcomes (e.g., economic downturn during a job-training study).
3. **Maturation**: Participants might change over time simply due to aging, learning, fatigue, etc., independent of the treatment.
4. **Testing Effects**: Taking a test more than once can influence participants' responses (practice effect).
5. **Instrumentation**: Changes in the measurement instrument or the observers can lead to inconsistencies in data collection.
6. **Regression to the Mean**: Extreme pre-test scores tend to move closer to the average on subsequent tests.
7. **Attrition (Mortality)**: Participants dropping out of the study in ways that are systematically related to the treatment or outcomes.

2.5.4.2 Strategies to Improve Internal Validity

- **Random Assignment**: Ensures that, on average, groups are equivalent on both known and unknown variables.
- **Control Groups**: Provide a baseline for comparison to isolate the effect of the intervention.
- **Blinding (Single-, Double-, or Triple-blind)**: Reduces biases from participants, researchers, or analysts.
- **Standardized Procedures and Protocols**: Minimizes variability in how interventions or measurements are administered.
- **Matching or Stratification**: When randomization is not possible, matching participants on key characteristics can reduce selection bias.
- **Pre-test–Post-test Designs**: Compare participant performance before and after the intervention (though watch for testing effects).

2.5.5 *External Validity*

External validity addresses the generalizability of the findings beyond the specific context of the study. A study with high external validity can be applied to other populations, settings, or times. On the other hand, localness can affect external validity.

2.5.5.1 Subtypes (or Related Concepts) of External Validity

1. **Population Validity**: The degree to which study findings can be generalized to the larger population from which the sample was drawn.
2. **Ecological Validity** (sometimes considered separately): Whether findings obtained in controlled conditions can be applied to real-world settings.
3. **Temporal Validity**: Whether the results of the study hold true over time. Changing societal norms, technologies, or economic conditions might render findings obsolete.

2.5.5.2 Threats to External Validity

- **Unrepresentative Samples**: If the sample does not reflect the wider population (in demographics, culture, etc.), generalization is limited.
- **Artificial Research Environments**: Highly controlled lab settings may not capture real-world complexities.
- **Treatment-Setting Interaction**: The effect of the treatment might depend on the unique conditions of the setting (e.g., a particular school, hospital, or region).
- **Treatment-Selection Interaction**: Certain characteristics of the selected participants might interact with the treatment (e.g., results from a specialized population do not apply to the general public).

2.5.5.3 Strategies to Improve External Validity

- **Use of Diverse and Representative Samples**: Recruit participants that mirror the larger population.
- **Field Studies and Naturalistic Settings**: Conduct experiments in real-world environments rather than artificial labs.
- **Replication in Multiple Contexts**: Replicate the study across different settings, geographic locations, and populations.
- **Longitudinal Studies**: Evaluate whether relationships hold over extended periods.

2.5.6 *Ecological Validity*

Ecological validity is often discussed as a subcategory of external validity. It specifically focuses on the *realism* of the study environment and tasks.

- **Definition**: The degree to which study findings can be generalized to the real-life settings where people actually live, work, and interact.
- **Key Idea**: Even if a lab experiment shows a particular behavior, do people behave the same way in their daily lives with everyday distractions, social pressures, and contextual factors?

2.5.6.1 Enhancing Ecological Validity

- **Naturalistic Observation**: Conduct observations or experiments in participants' usual environments.
- **Realistic Tasks**: Use tasks that closely mimic real-world challenges or behaviors.
- **Minimal Interference**: Researchers strive to reduce the artificiality of the setting, allowing participants to behave as naturally as possible.

2.5.7 *Statistical Conclusion Validity*

Though often discussed alongside internal validity, **statistical conclusion validity** focuses on whether the statistical tests used in a study are appropriate, powerful enough, and applied correctly.

2.5.7.1 Threats to Statistical Conclusion Validity

1. **Low Statistical Power**: If the sample size is too small, the study may fail to detect a real effect (Type II error).
2. **Violations of Statistical Assumptions**: Incorrect application of statistical tests can lead to spurious conclusions (e.g., using parametric tests with non-normal data without appropriate adjustments).
3. **Fishing and Error Rate Problem**: Running many statistical tests without correction increases the chance of a Type I error (finding a false positive).
4. **Reliability of Measures**: If the measurement instruments are unreliable, statistical correlations or differences may be undervalued or overstated.

2.5.7.2 Improving Statistical Conclusion Validity

- **Adequate Sample Size**: Conduct a power analysis to determine the necessary size to detect meaningful effects.
- **Appropriate Statistical Techniques**: Ensure your chosen analysis matches the nature of the data and research question.
- **Multiple Testing Corrections**: Use methods like Bonferroni or false discovery rate corrections when conducting multiple comparisons.
- **High-Quality Measurements**: Use reliable and valid measures to reduce measurement error.

2.5.8 *Putting It All Together*

1. **Face Validity**: Does it look like it measures what it should?
2. **Content Validity**: Does it cover all facets of the construct?
3. **Construct Validity**: Does it truly reflect the theoretical concept?
4. **Criterion Validity**: Does it correlate with or predict other relevant outcomes?
5. **Internal Validity**: Is the relationship between treatment and outcome truly causal?
6. **External Validity**: Can findings be generalized to other populations, settings, and times?
7. **Ecological Validity**: Are the findings applicable to real-world scenarios?
8. **Statistical Conclusion Validity**: Are the statistical inferences correct and robust?

Researchers typically need to strike a balance among these different validities:

- A **highly controlled lab study** might excel in internal validity but might have limited external and ecological validity.
- A **broad, naturalistic field study** might have stronger external or ecological validity but weaker internal validity due to less control over confounding variables.

No single study can maximize all validity types simultaneously, so replication, triangulation (using multiple methods), and transparent reporting are crucial strategies to bolster overall credibility.

2.6 Types of Subjects in a Treatment Setting

When conducting causal inference, particularly in randomized experiments or quasi-experimental settings, individuals in the study can be classified into four distinct groups based on their response to treatment assignment. These groups differ in how they react when they are assigned to receive or not receive treatment.

2.6.1 Non-switchers

Non-switchers are individuals whose treatment status does not change regardless of whether they are assigned to the treatment or control group. These individuals do not provide useful causal information because their behavior remains unchanged. They are further divided into:

- **Always-Takers**: These individuals will **always receive** the treatment, even if they are assigned to the control group.
- **Never-Takers**: These individuals will **never receive** the treatment, even if they are assigned to the treatment group.

Since their behavior is independent of the assignment, always-takers and never-takers do not contribute to identifying causal effects in standard randomized experiments. Instead, their presence can introduce bias in treatment effect estimation, particularly in **intention-to-treat analysis**.

2.6.2 Switchers

Switchers are individuals whose treatment status **depends on the assignment**. These individuals are the primary focus of causal inference because they provide meaningful information about the effect of treatment. They are classified into:

- **Compliers**: Individuals who follow the assigned treatment protocol.
  - If assigned to the treatment group, they **accept and receive** the treatment.
  - If assigned to the control group, they **do not receive** the treatment.
  - **Why are compliers important?**
    * They are the only group for whom treatment assignment affects actual treatment receipt.
    * Causal effect estimates (such as the local average treatment effect, LATE) are typically identified using compliers.
    * If the dataset only contains compliers, then the intention-to-treat effect (ITT) is equal to the treatment effect.

- **Defiers**: Individuals who do the **opposite** of what they are assigned.
  - If assigned to the treatment group, they **refuse the treatment**.
  - If assigned to the control group, they **seek out and receive** the treatment anyway.
  - **Why are defiers typically ignored?**
    * In most studies, defiers are assumed to be a small or negligible group.
    * Standard causal inference frameworks assume **monotonicity**, meaning no one behaves as a defier.

Table 2.2 Classification of Individuals Based on Treatment Assignment

	Treatment assignment	Control assignment
Compliers	Treated	Not treated
Always-takers	Treated	Treated
Never-takers	Not treated	Not treated
Defiers	Not treated	Treated

* If defiers exist in large numbers, estimating causal effects becomes significantly more complex.

2.6.3 *Classification of Individuals Based on Treatment Assignment*

The following Table 2.2 summarizes how different types of individuals respond to treatment and control assignments:

Key Takeaways

1. **Compliers** are the only group that allows us to estimate causal effects using **randomized or quasi-experimental designs**.
2. **Always-Takers and Never-Takers** do not provide meaningful variation in treatment status, making them less useful for causal inference.
3. **Defiers** typically violate the assumption of monotonicity, and their presence complicates causal estimation.
4. If a dataset consists **only of compliers**, the **intention-to-treat effect** will be equal to the **treatment effect**.

By correctly identifying and accounting for these different subject types, researchers can ensure more accurate causal inference and minimize biases in estimating treatment effects.

2.7 Types of Treatment Effects

When evaluating the causal impact of an intervention, different estimands (quantities of interest) can be used to measure treatment effects, depending on the study design and assumptions about compliance.

Terminology

- **Estimands**: The causal effect parameters we seek to measure
- **Estimators**: The statistical procedures used to estimate those parameters

- **Sources of Bias** (Keele & Grieve, 2025):

$$\begin{aligned}&\text{Estimator - True Causal Effect}\\&= \underbrace{\textbf{Hidden bias}}_{\text{Due to design}} + \underbrace{\textbf{Misspecification bias}}_{\text{Due to modeling}} + \underbrace{\textbf{Statistical noise}}_{\text{Due to finite sample}}\end{aligned}$$

1. **Hidden Bias (Due to Design)**
   - Arises from **unobserved confounders** and **measurement error** that remain after conditioning on observed covariates.
   - Is "hidden" because its true magnitude or direction cannot be directly observed.
   - Violations of **conditional exchangeability** (also called no unobserved confounding) imply the presence of hidden bias.
2. **Misspecification Bias (Due to Modeling)**
   - Occurs when the assumed model for the outcome or treatment assignment does not reflect the true data-generating process.
   - Persists even if we have perfect exchangeability (i.e., no hidden bias).
   - Can be viewed as **under-specification** (omitting essential terms or functional forms) or **over-specification** (including unnecessary parameters).
3. **Statistical Noise (Due to Finite Sample)**
   - Even with perfect design and correct model specification, finite samples lead to randomness in estimates.
   - Standard errors, confidence intervals, and p-values reflect this uncertainty.

In practice, all three sources of bias and uncertainty can coexist to varying degrees.

2.7.1 Average Treatment Effect

The Average Treatment Effect (ATE) is the expected difference in outcomes between individuals who receive treatment and those who do not.

Definition
Let:

- $Y_i(1)$ be the outcome of individual i under treatment.
- $Y_i(0)$ be the outcome of individual i under control.

The **individual treatment effect** is:

$$\tau_i = Y_i(1) - Y_i(0)$$

Since we cannot observe both $Y_i(1)$ and $Y_i(0)$ for the same individual (a fundamental problem in causal inference), we estimate the ATE across a population:

$$ATE = E[Y(1)] - E[Y(0)]$$

Identification Under Randomization

If treatment assignment is randomized (under Experimental Design), then the observed difference in means between treatment and control groups provides an unbiased estimator of ATE:

$$ATE = \frac{1}{N}\sum_{i=1}^{N}\tau_i = \frac{\sum_1^N Y_i(1)}{N} - \frac{\sum_i^N Y_i(0)}{N}$$

With **randomization**, we assume:

$$E[Y(1)|D = 1] = E[Y(1)|D = 0] = E[Y(1)]$$

$$E[Y(0)|D = 1] = E[Y(0)|D = 0] = E[Y(0)]$$

Thus, the difference in observed means between treated and control groups provides an unbiased estimate of ATE.

$$ATE = E[Y(1)] - E[Y(0)]$$

Alternatively, we can express the potential outcomes framework in a regression form, which allows us to connect causal inference concepts with standard regression analysis.

Instead of writing treatment effects as potential outcomes, we can define the observed outcome Y_i in terms of a regression equation:

$$Y_i = Y_i(0) + [Y_i(1) - Y_i(0)]D_i$$

where:

- $Y_i(0)$ is the outcome if individual i does not receive treatment.
- $Y_i(1)$ is the outcome if individual i does receive treatment.
- D_i is a binary indicator for treatment assignment.
- $D_i = 1$ if individual i receives treatment.
- $D_i = 0$ if individual i is in the control group.

We can redefine this equation using regression notation:

$$Y_i = \beta_{0i} + \beta_{1i} D_i$$

where:

- $\beta_{0i} = Y_i(0)$ represents the baseline (control group) outcome.
- $\beta_{1i} = Y_i(1) - Y_i(0)$ represents the individual treatment effect.

Thus, in an ideal setting, the coefficient on D_i in a regression gives us the treatment effect.

In observational studies, treatment assignment D_i is often **not random**, leading to **endogeneity**. This means that the error term in the regression equation might be correlated with D_i, violating one of the key assumptions of the [Ordinary Least Squares] estimator.

To formalize this issue, we can express the outcome equation as:

$$\begin{aligned}
Y_i &= \beta_{0i} + \beta_{1i} D_i \\
&= (\bar{\beta}_0 + \epsilon_{0i}) + (\bar{\beta}_1 + \epsilon_{1i}) D_i \\
&= \bar{\beta}_0 + \epsilon_{0i} + \bar{\beta}_1 D_i + \epsilon_{1i} D_i
\end{aligned}$$

where:

- $\bar{\beta}_0$ is the average baseline outcome.
- $\bar{\beta}_1$ is the average treatment effect.
- ϵ_{0i} captures individual-specific deviations in control group outcomes.
- ϵ_{1i} captures heterogeneous treatment effects.

If treatment assignment is truly **random**, then:

$$E[\epsilon_{0i}] = E[\epsilon_{1i}] = 0$$

which ensures:

- **No selection bias**: $D_i \perp \epsilon_{0i}$ (i.e., treatment assignment is independent of the baseline error).
- **Treatment effect is independent of assignment**: $D_i \perp \epsilon_{1i}$.

However, in observational studies, these assumptions often fail. This leads to:

- **Selection bias**: If individuals self-select into treatment based on unobserved characteristics, then D_i correlates with ϵ_{0i}.
- **Heterogeneous treatment effects**: If the treatment effect itself varies across individuals, then D_i correlates with ϵ_{1i}.

These issues violate the exogeneity assumption in OLS regression, leading to biased estimates of β_1.

When estimating treatment effects using OLS regression, we need to be aware of potential estimation issues.

1. **OLS Estimator and Difference-in-Means**

Under random assignment, the OLS estimator for β_1 simplifies to the difference in means estimator:

$$\hat{\beta}_1^{OLS} = \bar{Y}_{\text{treated}} - \bar{Y}_{\text{control}}$$

which is an unbiased estimator of the Average Treatment Effect.

However, when treatment assignment is not random, OLS estimates may be biased due to unobserved confounders.

2. **Heteroskedasticity and Robust Standard Errors**

If treatment effects vary across individuals (i.e., treatment effect heterogeneity), the error term contains an interaction:

$$\epsilon_i = \epsilon_{0i} + D_i \epsilon_{1i}$$

which leads to heteroskedasticity (i.e., the variance of errors depends on D_i and possibly on covariates X_i).

To address this, we use heteroskedasticity-robust standard errors, which ensure valid inference even when variance is not constant across observations.

2.7.2 *Conditional Average Treatment Effect*

Treatment effects may vary across different subgroups in a population. The Conditional Average Treatment Effect (CATE) captures heterogeneity in treatment effects across subpopulations.

For a subgroup characterized by covariates X_i:

$$CATE = E[Y(1) - Y(0)|X_i]$$

Why Is CATE Useful?

- **Heterogeneous Treatment Effects**: Certain groups may benefit more from treatment than others.
- **Policy Targeting**: Understanding who benefits the most allows for better resource allocation.

Example

- Policy Intervention: A job training program may have different effects on younger vs. older workers.
- Medical Treatments: Drug effectiveness may differ by gender, age, or genetic factors.

Estimating CATE allows policymakers and researchers to **identify who benefits most** from an intervention.

2.7.3 *Intention-to-Treat Effect*

A key issue in empirical research is **noncompliance**, where individuals do not always follow their assigned treatment (i.e., either people who are supposed to receive treatment don't receive it, or people who are supposed to be in the control group receive the treatment). The **Intention-to-Treat (ITT) effect** measures the impact of offering treatment, regardless of whether individuals actually receive it.

Definition
The ITT effect is the observed difference in means between groups assigned to treatment and control:

$$ITT = E[Y|D = 1] - E[Y|D = 0]$$

Why Use ITT?
- **Policy Evaluation**: ITT reflects the real-world effectiveness of an intervention, accounting for incomplete take-up.
- **Randomized Trials**: ITT preserves randomization, even when compliance is imperfect.

Example: Vaccination
- A government offers a vaccine (ITT), but not everyone actually takes it.
- The true treatment effect depends on those who receive the vaccine, which differs from the effect measured under ITT.

Since noncompliance is common in real-world settings, ITT effects are often smaller than true treatment effects. In this case, the difference in observed means between the treatment and control groups is not [Average Treatment Effects] but Intention-to-Treat Effect.

2.7.4 *Local Average Treatment Effects*

In many empirical settings, not all individuals assigned to treatment actually receive it (**noncompliance**). Instead of estimating the treatment effect for everyone assigned to treatment (i.e., Intention-to-Treat Effects), we often want to estimate the effect of treatment on those who actually **comply** with their assignment.

This is known as the Local Average Treatment Effect, also referred to as the **Complier Average Causal Effect (CACE)**.

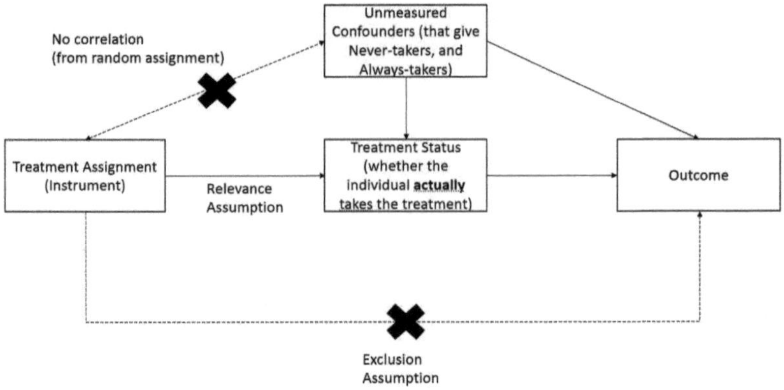

Fig. 2.2 LATE instrumental variables

- LATE is the treatment effect for the **subgroup of compliers**—those who take the treatment **if and only if** assigned to it.
- Unlike Conditional Average Treatment Effects, which describes heterogeneity across observable subgroups, **LATE focuses on compliance behavior**.
- We typically recover LATE using [Instrumental Variables], leveraging random treatment assignment as an instrument.

2.7.4.1 Estimating LATE Using Instrumental Variables

Instrumental variable estimation allows us to isolate the effect of treatment on **compliers** by using **random treatment assignment as an instrument for actual treatment receipt** (Fig. 2.2).

From an instrumental variables perspective (Table 2.3), LATE is estimated as:

$$LATE = \frac{ITT}{\text{Share of Compliers}}$$

where:

- **ITT (Intention-to-Treat Effect)** is the effect of being assigned to treatment.
- **Share of Compliers** is the proportion of individuals who actually take the treatment when assigned to it.

2.7.4.2 Key Properties of LATE

- As the proportion of compliers increases, LATE converges to ITT.

Table 2.3 Comparison of estimands under treatment noncompliance

Scenario	What it measures	When to use It?	Key assumptions
Intention-to-Treat	Effect of being assigned to treatment	Policy impact with noncompliance	None (preserves randomization)
LATE	Effect on compliers only	When we care about actual treatment effect rather than assignment	Excludability, monotonicity (no defiers)

- LATE is always larger than ITT, since ITT averages over both compliers and non-compliers.
- Standard error rule of thumb:
  - The standard error of LATE is given by:

$$SE(LATE) = \frac{SE(ITT)}{\text{Share of Compliers}}$$

- LATE can also be estimated using a pure placebo group (Gerber et al., 2010).
- Partial compliance is difficult to study.
- The IV/2SLS estimator is biased in small samples, requiring Bayesian methods for correction (Long et al., 2010; Jin & Rubin, 2009, 2008).

2.7.4.3 One-Sided Noncompliance

One-sided noncompliance occurs when we observe only compliers and never-takers in the sample (i.e., no always-takers).

Key Assumptions

- **Exclusion Restriction (Excludability)**: Never-takers have the same outcomes regardless of assignment (i.e., treatment has no effect on them because they never receive it).
- **Random Assignment Ensures Balance**: The number of never-takers is expected to be equal in the treatment and control groups.
  Estimation of LATE under one-sided noncompliance:

$$LATE = \frac{ITT}{\text{Share of Compliers}}$$

  Since the never-takers do not receive treatment, this simplifies estimation.

2.7.4.4 Two-Sided Noncompliance

Two-sided noncompliance occurs when we observe **compliers, never-takers, and always-takers** in the sample.

Key Assumptions

- **Exclusion Restriction (Excludability)**: Never-takers and always-takers have the same outcome **regardless of treatment assignment**.
- **Monotonicity Assumption** (No Defiers):
  - There are no defiers, meaning no individuals systematically avoid treatment when assigned to it.
  - This assumption is standard in practical studies.

  Estimation of LATE under two-sided noncompliance:

$$LATE = \frac{ITT}{\text{Share of Compliers}}$$

- Since always-takers receive treatment regardless of assignment, their presence does not bias LATE as long as monotonicity holds.
- In practice, monotonicity is often reasonable, as defiers are rare.

2.7.5 Population Versus Sample Average Treatment Effects

In experimental and observational studies, we often estimate the **Sample Average Treatment Effect (SATE)** using a finite sample. However, the **Population Average Treatment Effect (PATE)** is the parameter of interest when making broader generalizations.

Key Issue

SATE **does not necessarily equal** PATE due to **sample selection bias and treatment imbalance**.

See Imai et al. (2008) for an in-depth discussion on when SATE diverges from PATE.

Consider a **finite population** of size N from which we observe a sample of size n ($N \gg n$). Half of the sample receives treatment, and half is assigned to control.

Define the following indicators:

- **Sampling Indicator**:

$$I_i = \begin{cases} 1, & \text{if unit } i \text{ is in the sample} \\ 0, & \text{otherwise} \end{cases}$$

- **Treatment Assignment Indicator**:

$$T_i = \begin{cases} 1, & \text{if unit } i \text{ is in the treatment group} \\ 0, & \text{if unit } i \text{ is in the control group} \end{cases}$$

- **Potential Outcomes Framework**:

$$Y_i = \begin{cases} Y_i(1), & \text{if } T_i = 1 \text{ (Treated)} \\ Y_i(0), & \text{if } T_i = 0 \text{ (Control)} \end{cases}$$

- **Observed Outcome**:
  Since we can never observe both potential outcomes for the same unit, the observed outcome is:

$$Y_i | I_i = 1 = T_i Y_i(1) + (1 - T_i) Y_i(0)$$

- **True Individual Treatment Effect**:
  The individual-level treatment effect is:

$$TE_i = Y_i(1) - Y_i(0)$$

However, since we observe only one of $Y_i(1)$ or $Y_i(0)$, TE_i is never directly observed.

2.7.5.1 Definitions of SATE and PATE

- **Sample Average Treatment Effect (SATE):**

$$SATE = \frac{1}{n} \sum_{i \in \{I_i = 1\}} TE_i$$

  SATE is the **average treatment effect within the sample**.
- **Population Average Treatment Effect (PATE):**

$$PATE = \frac{1}{N} \sum_{i=1}^{N} TE_i$$

  PATE represents the **true treatment effect across the entire population**.

  Since we observe only a subset of the population, SATE **may not equal** PATE.

2.7.5.2 Decomposing Estimation Error

The baseline estimator for SATE and PATE is the **difference in observed means**:

$$D = \frac{1}{n/2} \sum_{i \in (I_i=1, T_i=1)} Y_i - \frac{1}{n/2} \sum_{i \in (I_i=1, T_i=0)} Y_i$$

$$= \text{(Mean of Treated Group)} - \text{(Mean of Control Group)}$$

Define Δ as the estimation error (i.e., deviation from the truth), under an additive model:

$$Y_i(t) = g_t(X_i) + h_t(U_i)$$

The estimation error is decomposed into

$$\begin{aligned} PATE - D = \Delta = {} & \Delta_S + \Delta_T \\ & = (PATE - SATE) + (SATE - D) \\ & = \text{Sample Selection Bias} + \text{Treatment Imbalance} \\ & = (\Delta_{S_X} + \Delta_{S_U}) + (\Delta_{T_X} + \Delta_{T_U}) \\ & = \text{(Selection on Observables + Selection on Unobservables)} \\ & + \text{(Treatment Imbalance in Observables} \\ & + \text{Treatment Imbalance in Unobservables)} \end{aligned}$$

To further illustrate this, we begin by explicitly defining how the total discrepancy $PATE - D$ separates into different components.

Step 1: From $PATE - D$ to $\Delta_S + \Delta_T$

$$\underbrace{PATE - D}_{\Delta} = \underbrace{(PATE - SATE)}_{\Delta_S} + \underbrace{(SATE - D)}_{\Delta_T}.$$

- $PATE - D$: The total discrepancy between the true population treatment effect and the estimate D
- $\Delta_S = PATE - SATE$: Sample Selection Bias—how much the sample ATE differs from the population ATE
- $\Delta_T = SATE - D$: Treatment Imbalance—how much the estimated treatment effect deviates from the sample ATE

Step 2: Breaking Bias into Observables and Unobservables

Each bias term can be decomposed into observed (X) and unobserved (U) factors:

$$\Delta_S = \underbrace{\Delta_{S_X}}_{\text{Selection on Observables}} + \underbrace{\Delta_{S_U}}_{\text{Selection on Unobservables}}$$

$$\Delta_T = \underbrace{\Delta_{T_X}}_{\text{Treatment Imbalance in Observables}} + \underbrace{\Delta_{T_U}}_{\text{Treatment Imbalance in Unobservables}}$$

Thus, the final expression:

$$\begin{aligned} PATE - D &= \underbrace{(PATE - SATE)}_{\Delta_S\text{: Sample Selection Bias}} + \underbrace{(SATE - D)}_{\Delta_T\text{: Treatment Imbalance}} \\ &= \underbrace{(\Delta_{S_X} + \Delta_{S_U})}_{\text{Selection on } X + \text{ Selection on } U} + \underbrace{(\Delta_{T_X} + \Delta_{T_U})}_{\text{Imbalance in } X + \text{ Imbalance in } U} . \end{aligned}$$

This decomposition clarifies the sources of error in estimating the true effect, distinguishing between sample representativeness (selection bias) and treatment assignment differences (treatment imbalance), and further separating these into observable and unobservable components.

2.7.5.2.1 Sample Selection Bias (Δ_S)

Also called sample selection error, this arises when the sample is not representative of the population.

$$\Delta_S = PATE - SATE = \frac{N-n}{N}(NATE - SATE)$$

where:

- NATE (Non-sample Average Treatment Effect) is the average treatment effect for the part of the population not included in the sample:

$$NATE = \sum_{i \in (I_i = 0)} \frac{TE_i}{N-n}$$

To eliminate sample selection bias ($\Delta_S = 0$):

1. Redefine the sample as the entire population ($N = n$).
2. Ensure $NATE = SATE$ (e.g., treatment effects must be homogeneous across sampled and non-sampled units).

However, when treatment effects vary across individuals, random sampling only warrants sample selection bias but does not sample eliminate error.

2.7.5.2.2 Treatment Imbalance Error (Δ_T)

Also called treatment imbalance bias, this occurs when the empirical distribution of treated and control units differs.

$$\Delta_T = SATE - D$$

Key Insight
$\Delta_T \rightarrow 0$ when the treatment and control groups are balanced across both observables (X) and unobservables (U).

Since we cannot directly adjust for unobservables, imbalance correction methods focus on observables.

2.7.5.3 Adjusting for (Observable) Treatment Imbalance

However, in real-world studies:

- We can only adjust for observables X, not unobservables U.
- Residual imbalance in unobservables may still introduce bias after adjustment (Table 2.4).

To address treatment imbalance, researchers commonly use:

1. Blocking
2. [Matching Methods]

2.7.6 *Average Treatment Effects on the Treated and Control*

In many empirical studies, researchers are interested in how treatment affects **specific subpopulations** rather than the entire population. Two commonly used treatment effect measures are:

1. **Average Treatment Effect on the Treated** (ATT): The effect of treatment on individuals who actually received treatment.
2. **Average Treatment Effect on the Control** (ATC): The effect treatment would have had on individuals who were not treated.

Understanding the distinction between ATT, ATC, and ATE is crucial for determining external validity and for designing targeted policies.

Table 2.4 Comparison of blocking and matching methods for covariate balance

Method	Blocking	[Matching methods]
Definition	Random assignment within predefined strata based on pretreatment covariates	Dropping, repeating, or grouping observations to balance covariates between treated and control groups (Rubin, 1973)
When applied?	Before treatment assignment (in experimental designs)	After treatment assignment (in observational studies)
Effectiveness	Ensures exact balance within strata but may require large sample sizes for fine stratification	Can improve balance, but risk of increasing bias if covariates are poorly chosen
What if covariates are irrelevant?	No effect on treatment estimates	Worst-case scenario: If matching is done on covariates uncorrelated with treatment but correlated with outcomes, it may increase bias instead of reducing it
Benefits	Eliminates imbalance in observables ($\Delta_{T_X} = 0$). Effect on unobservables is uncertain (may help if unobservables correlate with observables)	Reduces model dependence, bias, variance, and mean-squared error (MSE). Matching only balances observables, and its effect on unobservables is unknown

2.7.6.1 Average Treatment Effect on the Treated

The **ATT** measures the expected treatment effect **only for those who were actually treated**:

$$\begin{aligned} ATT &= E[Y_i(1) - Y_i(0)|D_i = 1] \\ &= E[Y_i(1)|D_i = 1] - E[Y_i(0)|D_i = 1] \end{aligned}$$

Key Interpretation

- ATT tells us how much better (or worse) off treated individuals are compared to their hypothetical counterfactual outcome (had they not been treated).
- It is useful for evaluating the effectiveness of interventions on those who self-select into treatment.

2.7.6.2 Average Treatment Effect on the Control

The **ATC** measures the expected treatment effect **only for those who were not treated**:

$$\begin{aligned} ATC &= E[Y_i(1) - Y_i(0)|D_i = 0] \\ &= E[Y_i(1)|D_i = 0] - E[Y_i(0)|D_i = 0] \end{aligned}$$

Key Interpretation

- ATC answers the question: "What would have been the effect of treatment if it had been given to those who were not treated?"
- It is important for understanding how an intervention might generalize to untreated populations.

2.7.6.3 Relationship Between ATT, ATC, and ATE

Under **random assignment and full compliance**, we have:

$$ATE = ATT = ATC$$

Why?

- Randomization ensures that treated and untreated groups are statistically identical before treatment.
- Thus, treatment effects are the same across groups, leading to ATT = ATC = ATE.

However, in observational settings, selection bias and treatment heterogeneity may cause ATT and ATC to diverge from ATE.

2.7.6.4 Sample Average Treatment Effect on the Treated

The **Sample ATT (SATT)** is the empirical estimate of ATT in a finite sample:

$$SATT = \frac{1}{n} \sum_{i \in D_i = 1} TE_i$$

where:

- $TE_i = Y_i(1) - Y_i(0)$ is the treatment effect for unit i.
- n is the number of treated units in the sample.
- The summation is taken only over treated units in the sample.

2.7.6.5 Population Average Treatment Effect on the Treated

The **Population ATT (PATT)** generalizes ATT to the entire treated population:

$$PATT = \frac{1}{N} \sum_{i \in D_i = 1} TE_i$$

where:

- $TE_i = Y_i(1) - Y_i(0)$ is the treatment effect for unit i.

- N is the total number of treated units in the population.
- The summation is taken over all treated individuals in the population.

If the sample is randomly drawn, then $SATT \approx PATT$, but if the sample is not representative, $SATT$ may overestimate or underestimate $PATT$.

2.7.6.6 When ATT and ATC Diverge from ATE

In real-world studies, ATT and ATC often differ from ATE due to treatment effect heterogeneity and selection bias.

2.7.6.6.1 Selection Bias in ATT

If individuals self-select into treatment, then the treated group may be systematically different from the control group.

- **Example**:
  - Suppose a job training program is voluntary.
  - Individuals who enroll might be more motivated or have better skills than those who do not.
  - As a result, the treatment effect (ATT) may not generalize to the untreated group (ATC).

  This implies:

$$ATT \neq ATC$$

  unless treatment assignment is random.

2.7.6.6.2 Treatment Effect Heterogeneity

If treatment effects vary across individuals, then:

- ATT may be larger or smaller than ATE, depending on how treatment effects differ across subgroups.
- ATC may be larger or smaller than ATT, if the untreated group would have responded differently to treatment.

Example
- A scholarship program may be more beneficial for students from lower-income families than for students from wealthier backgrounds.
- If lower-income students are more likely to apply for the scholarship, then ATT > ATE.

Table 2.5 Comparison of treatment effect estimands: ATE, ATT, and ATC

Treatment effect	Definition	Use case	Potential issues
ATE (Average Treatment Effect)	Effect on randomly selected individuals	Policy decisions applicable to entire population	Requires full randomization
ATT (Average Treatment on Treated)	Effect on those who received treatment	Evaluating effectiveness of interventions for targeted groups	Selection bias if treatment is voluntary
ATC (Average Treatment on Control)	Effect if treatment were given to untreated individuals	Predicting treatment effects for new populations	May not be generalizable

- However, if wealthier students (who did not receive the scholarship) would have benefited less from it, then ATC < ATE (Table 2.5).
  Thus, we may observe:

$$ATE \neq ATT \neq ATC$$

2.7.7 *Quantile Average Treatment Effects*

Instead of focusing on the mean effect (ATE), Quantile Treatment Effects (QTE) help us understand how treatment shifts the entire distribution of an outcome variable.

The Quantile Treatment Effect at quantile τ is defined as:

$$QTE_\tau = Q_\tau(Y_1) - Q_\tau(Y_0)$$

where:

- $Q_\tau(Y_1)$ is the τ-th quantile of the outcome distribution **under treatment**.
- $Q_\tau(Y_0)$ is the τ-th quantile of the outcome distribution **under control**.

When to Use QTE?

- **Heterogeneous Treatment Effects**: If treatment effects differ across individuals, ATE may be misleading.
- **Policy Targeting**: Policymakers may care more about low-income individuals (e.g., bottom 25%) rather than the average effect.
- **Distributional Changes**: QTE allows us to assess whether treatment increases inequality (e.g., benefits the rich more than the poor).

Estimation of QTE
QTE can be estimated using:

- **Quantile Regression**: Extends linear regression to estimate effects at different quantiles.
- **[Instrumental Variables] for QTE**: Requires additional assumptions to estimate causal effects in the presence of endogeneity (Abadie et al., 2002; Chernozhukov & Hansen, 2005).

Example: Wage Policy Impact
- Suppose a minimum wage increase is introduced.
- The ATE might show a small positive effect on earnings.
- However, QTE might reveal:
  - No effect at the bottom quantiles (for workers who lose jobs).
  - A positive effect at the median.
  - A strong positive effect at the top quantiles (for experienced workers who benefit the most).

Thus, QTE provides a more detailed view of the treatment effect across the entire income distribution.

2.7.8 Log-Odds Treatment Effects for Binary Outcomes

When the **outcome variable is binary** (e.g., success/failure, employed/unemployed, survived/died), it is often useful to measure the treatment effect in **log-odds form**.

For a binary outcome Y, define the **probability of success** as:

$$P(Y = 1|D = d)$$

The **log-odds of success** under treatment and control are:

$$\text{Log-odds}(Y|D = 1) = \log\left(\frac{P(Y = 1|D = 1)}{1 - P(Y = 1|D = 1)}\right)$$

$$\text{Log-odds}(Y|D = 0) = \log\left(\frac{P(Y = 1|D = 0)}{1 - P(Y = 1|D = 0)}\right)$$

The **Log-Odds Treatment Effect (LOTE)** is then:

$$LOTE = \text{Log-odds}(Y|D = 1) - \text{Log-odds}(Y|D = 0)$$

This captures how treatment **affects the relative likelihood** of success in a nonlinear way.

When to Use Log-Odds Treatment Effects?

- **Binary Outcomes**: When the treatment outcome is 0 or 1 (e.g., employed/unemployed).
- **Nonlinear Treatment Effects**: Log-odds help handle situations where effects are multiplicative rather than additive.
- **Rare Events**: Useful in cases where the outcome probability is very small or very large.

Estimation of Log-Odds Treatment Effects

- **Logistic Regression with Treatment Indicator**:

$$\log\left(\frac{P(Y=1|D=1)}{1-P(Y=1|D=1)}\right)=\beta_0+\beta_1 D$$

  where β_1 represents the **log-odds treatment effect**.
- **Randomization-Based Estimation**: Freedman (2008) provides a framework for randomized trials that ensures consistent estimation.
- **Attributable Effects**: Alternative methods, such as those in Rosenbaum (2002), estimate the proportion of cases attributable to the treatment.

2.7.9 Summary Table: Treatment Effect Estimands

Table 2.6.

Table 2.6 Treatment effect estimands

Treatment effect	Definition	Use case	Key assumptions	When it differs from ATE?
Average Treatment Effect	The expected treatment effect for a randomly chosen individual in the population	General policy evaluation; measures the overall impact	Randomization or strong ignorability (treatment assignment independent of potential outcomes)	–
Conditional Average Treatment Effect	The treatment effect for a specific subgroup of the population, conditional on covariates *X*	Identifies heterogeneous effects; useful for targeted interventions	Treatment effect heterogeneity must exist	Differs when treatment effects vary across subgroups
Intention-to-Treat Effect	The effect of being assigned to treatment, regardless of actual compliance	Policy evaluations where noncompliance exists	Randomized treatment assignment ensures unbiased estimation	Lower than ATE when not all assigned individuals comply
Local Average Treatment Effect	The effect of treatment only on compliers—those who take the treatment if and only if assigned to it	When compliance is imperfect, LATE isolates the effect for compliers	Monotonicity (no defiers); instrument only affects the outcome through treatment	Differs from ATE when compliance is selective
Average Treatment Effect on the Treated	The effect of treatment on those who actually received the treatment	Used when assessing effectiveness of a treatment for those who self-select into it	No unmeasured confounders within the treated group.	Differs when treatment selection is not random.
Average Treatment Effect on the Control	The effect the treatment would have had on individuals who were not treated	Predicts the effect of expanding a program to the untreated population.	No unmeasured confounders within the control group	Differs when treatment effects are heterogeneous

(continued)

Table 2.6 (continued)

Treatment effect	Definition	Use case	Key assumptions	When it differs from ATE?
Sample Average Treatment Effect	The estimated treatment effect in the sample	Used when evaluating treatment within a specific sample	Sample must be representative of the population for external validity	Differs when the sample is not representative of the population
Population Average Treatment Effect	The expected treatment effect for the entire population	Policy design and large-scale decision-making	Requires that sample selection is random	Differs when sample selection bias exists
Quantile Treatment Effect	The treatment effect at a specific percentile of the outcome distribution	Understanding distributional effects rather than mean effects	Rank preservation or monotonicity assumptions may be needed	Differs when treatment effects vary across outcome quantiles
Log-Odds Treatment Effect	The effect of treatment on binary outcomes, expressed in log-odds	Used when outcomes are dichotomous (e.g., employed/unemployed, survived/died)	Logistic model assumptions must hold	Differs when treatment effects are nonlinear or outcome probabilities are low

Chapter 3
Experimental Design

This chapter provides a treatment of experimental design principles, focusing on their application in business experiments such as A/B testing, consumer behavior studies, and market research. It begins by defining key terms and explaining why randomization is the gold standard for establishing causal relationships. The chapter covers both classical and advanced designs, including completely randomized designs, factorial designs, and crossover trials. It also delves into the selection problem and the consequences of bias. Topics such as blocking, stratification, and factorial manipulation are discussed with an emphasis on practical implementation. Statistical notation is introduced to describe treatment allocation mechanisms, and visual diagrams help clarify complex ideas. Emerging research trends and limitations of traditional designs are also explored to prepare readers for real-world complexities.

Imagine you're a marketing manager trying to decide whether a new advertising campaign will boost sales. Or perhaps you're an economist investigating the impact of tax incentives on consumer spending. In both cases, you need a way to determine **causal effects**—not just correlations. This is where **experimental design** becomes essential.

At its core, experimental design ensures that studies are conducted efficiently and that valid conclusions can be drawn. In business applications, experiments help quantify the effects of pricing strategies, marketing tactics, and financial interventions. A well-designed experiment reduces bias, controls variability, and maximizes the accuracy of causal inferences.

3.1 Principles of Experimental Design

A well-designed experiment follows three key principles:

1. **Randomization**: Ensures that subjects or experimental units are assigned to different groups randomly, eliminating selection bias

M. Nguyen, *Experimental Design*, https://doi.org/10.1007/978-3-032-01839-7_3

2. **Replication**: Increases the precision of estimates by repeating the experiment on multiple subjects
3. **Control**: Isolates the effect of treatments by using control groups or baseline conditions to minimize confounding factors

In addition to these, **blocking** and **factorial designs** further refine experimental accuracy.

3.2 The Gold Standard: Randomized Controlled Trials

Randomized Controlled Trials (RCTs) are the **holy grail of causal inference**. Their power comes from **random assignment**, which ensures that any differences between treatment and control groups arise **only** due to the intervention.
RCTs provide:

- **Unbiased estimates** of treatment effects
- **Elimination of confounding factors on average** (although covariate imbalance can occur, necessitating techniques like Re-randomization to achieve a "platinum standard" set by Tukey (1993))

An RCT consists of two groups:

1. **Treatment group**: Receives the intervention (e.g., a new marketing campaign, drug, or financial incentive)
2. **Control group**: Does not receive the intervention, serving as a baseline

Subjects from the same population are randomly assigned to either group. This randomization ensures that any observed differences in outcomes are due solely to the treatment—not external factors.

However, RCTs are easier to conduct in **hard sciences** (e.g., medicine or physics), where treatments and environments can be tightly controlled. In **social sciences**, challenges arise because:

- Human behavior is unpredictable.
- Some treatments are impossible or unethical to introduce (e.g., assigning individuals to poverty).
- Real-world environments are difficult to control.

To address these challenges, social scientists use Quasi-Experimental Methods to approximate experimental conditions.

RCTs establish internal validity, meaning that the observed treatment effects are **causally interpretable**. Even though **random assignment is not the same as *ceteris paribus*** (holding everything else constant), it achieves a similar effect: **on average**, treatment and control groups should be comparable.

3.3 Selection Problem

A fundamental challenge in causal inference is that we never observe both potential outcomes for the same individual—only one or the other. This creates the selection problem, which we formalize below.
Assume we have:

- A **binary treatment variable**:
  $D_i \in \{0, 1\}$, where:
  - $D_i = 1$ indicates that individual i receives the treatment.
  - $D_i = 0$ indicates that individual i does **not** receive the treatment.
- The **outcome of interest**:
  Y_i, which depends on whether the individual is treated or not:
  - Y_{0i}: The outcome **if not treated.**
  - Y_{1i}: The outcome **if treated.**

Thus, the **potential outcomes framework** is defined as:

$$\text{Potential Outcome} = \begin{cases} Y_{1i}, & \text{if } D_i = 1 \quad \text{(Treated)} \\ Y_{0i}, & \text{if } D_i = 0 \quad \text{(Untreated)} \end{cases}$$

However, we only **observe** one outcome per individual:

$$Y_i = Y_{0i} + (Y_{1i} - Y_{0i})D_i$$

This means that for any given person, we either observe Y_{1i} or Y_{0i}, but **never both.** Since we cannot observe counterfactuals (unless we invent a time machine), we must rely on statistical inference to estimate treatment effects.

3.3.1 The Observed Difference in Outcomes

The goal is to estimate the difference in expected outcomes between treated and untreated individuals:

$$E[Y_i|D_i = 1] - E[Y_i|D_i = 0]$$

Expanding this equation:

$$
\begin{aligned}
E[Y_i|D_i = 1] - E[Y_i|D_i = 0] &= (E[Y_{1i}|D_i = 1] - E[Y_{0i}|D_i = 1]) \\
&\quad + (E[Y_{0i}|D_i = 1] - E[Y_{0i}|D_i = 0]) \\
&= (E[Y_{1i} - Y_{0i}|D_i = 1]) \\
&\quad + (E[Y_{0i}|D_i = 1] - E[Y_{0i}|D_i = 0])
\end{aligned}
$$

This equation decomposes the **observed difference** into two components:

- Treatment Effect on the Treated: $E[Y_{1i} - Y_{0i}|D_i = 1]$, which represents the causal impact of the treatment on those who are treated.
- **Selection Bias**:
  $E[Y_{0i}|D_i = 1] - E[Y_{0i}|D_i = 0]$, which captures systematic differences between treated and untreated groups **even in the absence of treatment**.

Thus, the **observed difference in outcomes** is:

$$
\text{Observed Difference} = \text{ATT} + \text{Selection Bias}
$$

3.3.2 Eliminating Selection Bias with Random Assignment

With **random assignment** of treatment, D_i is independent of potential outcomes:

$$
E[Y_i|D_i = 1] - E[Y_i|D_i = 0] = E[Y_{1i} - Y_{0i}]
$$

This works because, under true randomization:

$$
E[Y_{0i}|D_i = 1] = E[Y_{0i}|D_i = 0]
$$

which eliminates selection bias. Consequently, the observed difference now **directly estimates the true causal effect**:

$$
E[Y_i|D_i = 1] - E[Y_i|D_i = 0] = E[Y_{1i} - Y_{0i}]
$$

Thus, randomized controlled trials provide an unbiased estimate of the average treatment effect.

3.3.3 *Another Representation Under Regression*

So far, we have framed the selection problem using expectations and potential outcomes. Another way to represent treatment effects is through **regression models**, which provide a practical framework for estimation.

Suppose the treatment effect is constant across individuals:

$$Y_{1i} - Y_{0i} = \rho$$

This implies that each treated individual experiences the same treatment effect (ρ), though their baseline outcomes (Y_{0i}) may vary.

Since we only observe **one** of the potential outcomes, the observed outcome can be expressed as:

$$\begin{aligned} Y_i &= E(Y_{0i}) + (Y_{1i} - Y_{0i})D_i + [Y_{0i} - E(Y_{0i})] \\ &= \alpha + \rho D_i + \eta_i \end{aligned}$$

where:

- $\alpha = E(Y_{0i})$, the expected outcome for untreated individuals.
- ρ represents the **causal treatment effect**.
- $\eta_i = Y_{0i} - E(Y_{0i})$, capturing individual deviations from the mean untreated outcome.

Thus, the regression model provides an intuitive way to express treatment effects.

3.3.3.1 Conditional Expectations and Selection Bias

Taking expectations conditional on treatment status:

$$\begin{aligned} E[Y_i|D_i = 1] &= \alpha + \rho + E[\eta_i|D_i = 1] \\ E[Y_i|D_i = 0] &= \alpha + E[\eta_i|D_i = 0] \end{aligned}$$

The observed difference in means between treated and untreated groups is:

$$E[Y_i|D_i = 1] - E[Y_i|D_i = 0] = \rho + E[\eta_i|D_i = 1] - E[\eta_i|D_i = 0]$$

Here, the term $E[\eta_i|D_i = 1] - E[\eta_i|D_i = 0]$ represents **selection bias**—the correlation between the regression error term (η_i) and the treatment variable (D_i).

Under random assignment, we assume that potential outcomes are **independent** of treatment (D_i):

$$E[\eta_i|D_i = 1] - E[\eta_i|D_i = 0] = E[Y_{0i}|D_i = 1] - E[Y_{0i}|D_i = 0] = 0$$

Thus, under true **randomization**, selection bias disappears, and the observed difference directly estimates the **causal effect** ρ.

3.3.3.2 Controlling for Additional Variables

In many real-world scenarios, **random assignment is imperfect**, and selection bias may still exist. To mitigate this, we introduce **control variables** (X_i), such as demographic characteristics, firm size, or prior purchasing behavior.

If X_i is **uncorrelated with treatment** (D_i), including it in our regression model does not bias the estimate of ρ but has two advantages:

1. **It reduces residual variance** (η_i), improving the precision of ρ.
2. **It accounts for additional sources of variability**, making the model more robust.

Thus, our regression model extends to:

$$Y_i = \alpha + \rho D_i + X_i' \gamma + \eta_i$$

where:

- X_i represents a vector of control variables.
- γ captures the effect of X_i on the outcome.

3.3.3.3 Example: Racial Discrimination in Hiring

A famous study by Bertrand and Mullainathan (2004) examined racial discrimination in hiring by randomly assigning Black- and White-sounding names to identical job applications. By ensuring that names were assigned randomly, the authors eliminated confounding factors like education and experience, allowing them to estimate the causal effect of race on callback rates.

3.4 Classical Experimental Designs

Experimental designs provide structured frameworks for conducting experiments, ensuring that results are **statistically valid** and **practically applicable**. The choice of design depends on the research question, the nature of the treatment, and potential sources of variability. For a more in-depth statistical understanding of these designs, we will revisit them again in Analysis of Variance.

3.4.1 Completely Randomized Design

In a Completely Randomized Design (CRD), each experimental unit is randomly assigned to a treatment group. This is the simplest form of experimental design and is effective when no confounding factors are present.

Example: Email Marketing Experiment
A company tests three different email marketing strategies (A, B, and C) to measure their effect on **customer engagement** (click-through rate). Customers are randomly assigned to receive one of the three emails.

Mathematical Model

$$Y_{ij} = \mu + \tau_i + \epsilon_{ij}$$

where:

- Y_{ij} is the response variable (e.g., click-through rate).
- μ is the overall mean response.
- τ_i is the effect of treatment i.
- ϵ_{ij} is the random error term, assumed to be normally distributed: $\epsilon_{ij} \sim N(0, \sigma^2)$.

```
set.seed(123)

# Simulated dataset for email marketing experiment
data <- data.frame(
  group = rep(c("A", "B", "C"), each = 10),
  response = c(rnorm(10, mean=50, sd=5),
               rnorm(10, mean=55, sd=5),
               rnorm(10, mean=60, sd=5))
)

# ANOVA to test for differences among groups
anova_result <- aov(response ~ group, data = data)
summary(anova_result)
#>             Df Sum Sq Mean Sq F value  Pr(>F)
#> group        2  306.1  153.04   6.435 0.00518 **
#> Residuals   27  642.1   23.78
#> ---
#> Signif. codes:  0 '***' 0.001 '**' 0.01 '*' 0.05 '.' 0.1 ' ' 1
```

If the **p-value** in the ANOVA summary is **less than 0.05**, we reject the null hypothesis and conclude that at least one email strategy significantly affects engagement.

3.4.2 Randomized Block Design

A Randomized Block Design (RBD) is used when experimental units can be grouped into homogeneous blocks based on a known confounding factor. Blocking helps reduce unwanted variation, increasing the precision of estimated treatment effects.

Example: Store Layout Experiment
A retailer tests three store layouts (A, B, and C) on sales performance. Since store location (Urban, Suburban, Rural) might influence sales, we use blocking to control for this effect.

Mathematical Model

$$Y_{ij} = \mu + \tau_i + \beta_j + \epsilon_{ij}$$

where:

- Y_{ij} is the sales outcome for store i in location j.
- μ is the overall mean sales.
- τ_i is the effect of layout i.
- β_j represents the block effect (location).
- ϵ_{ij} is the random error.

```
set.seed(123)

# Simulated dataset for store layout experiment
data <- data.frame(
  block = rep(c("Urban", "Suburban", "Rural"), each = 6),
  layout = rep(c("A", "B", "C"), times = 6),
  sales = c(rnorm(6, mean=200, sd=20),
            rnorm(6, mean=220, sd=20),
            rnorm(6, mean=210, sd=20))
)

# ANOVA with blocking factor
anova_block <- aov(sales ~ layout + block, data = data)
summary(anova_block)
#>             Df Sum Sq Mean Sq F value Pr(>F)
#> layout       2     71    35.7   0.071  0.931
#> block        2    328   164.1   0.328  0.726
#> Residuals   13   6500   500.0
```

By including `block` in the model, we account for **location effects**, leading to **more accurate** treatment comparisons.

3.4.3 Factorial Design

A Factorial Design evaluates the effects of two or more factors simultaneously, allowing for the study of interactions between variables.

Example: Pricing and Advertising Experiment
A company tests two pricing strategies (High, Low) and two advertising methods (TV, Social Media) on sales.

Mathematical Model

$$Y_{ijk} = \mu + \tau_i + \gamma_j + (\tau\gamma)_{ij} + \epsilon_{ijk}$$

where:

- τ_i is the effect of price level i.
- γ_j is the effect of advertising method j.
- $(\tau\gamma)_{ij}$ is the **interaction effect** between price and advertising.
- ϵ_{ijk} is the random error term.

```
set.seed(123)

# Simulated dataset
data <- expand.grid(
  Price = c("High", "Low"),
  Advertising = c("TV", "Social Media"),
  Replicate = 1:10
)

# Generate response variable (sales)
data$Sales <- with(data,
                   100 +
                   ifelse(Price == "Low", 10, 0) +
                   ifelse(Advertising == "Social Media", 15,
                   ↪  0) +
                   ifelse(Price == "Low" & Advertising ==
                   ↪  "Social Media", 5, 0) +
                   rnorm(nrow(data), sd=5))

# Two-way ANOVA
anova_factorial <- aov(Sales ~ Price * Advertising, data =
↪  data)
summary(anova_factorial)
#>                   Df Sum Sq Mean Sq F value   Pr(>F)
#> Price              1   1364    1364   66.60 1.05e-09 ***
```

```
#> Advertising        1   3640    3640  177.67 1.72e-15 ***
#> Price:Advertising  1     15      15    0.71    0.405
#> Residuals         36    738      20
#> ---
#> Signif. codes:  0 '***' 0.001 '**' 0.01 '*' 0.05 '.' 0.1 ' ' 1
```

3.4.4 Crossover Design

A Crossover Design is used when each subject receives multiple treatments in a sequential manner. This design controls for individual differences by using each subject as their own control.

Example: Drug Trial
Patients receive Drug A in the first period and Drug B in the second period, or vice versa.

Mathematical Model

$$Y_{ijk} = \mu + \tau_i + \pi_j + \beta_k + \epsilon_{ijk}$$

where:

- τ_i is the **treatment effect**.
- π_j is the **period effect** (e.g., learning effects).
- β_k is the **subject effect** (individual baseline differences).

```
set.seed(123)

# Simulated dataset
data <- data.frame(
  Subject = rep(1:10, each = 2),
  Period = rep(c("Period 1", "Period 2"), times = 10),
  Treatment = rep(c("A", "B"), each = 10),
  Response = c(rnorm(10, mean=50, sd=5), rnorm(10, mean=55,
  ↪  sd=5))
)

# Crossover ANOVA
anova_crossover <-
  aov(Response ~ Treatment + Period + Error(Subject / Period),
      data = data)
```

```
summary(anova_crossover)
#>
#> Error: Subject
#>           Df Sum Sq Mean Sq
#> Treatment  1  63.94   63.94
#>
#> Error: Subject:Period
#>        Df Sum Sq Mean Sq
#> Period  1  21.92   21.92
#>
#> Error: Within
#>           Df Sum Sq Mean Sq F value Pr(>F)
#> Treatment  1  134.9  134.91   5.231 0.0371 *
#> Period     1    0.2    0.24   0.009 0.9237
#> Residuals 15  386.9   25.79
#> ---
#> Signif. codes:  0 '***' 0.001 '**' 0.01 '*' 0.05 '.' 0.1 ' ' 1
```

3.4.5 Split-Plot Design

A Split-Plot Design is used when one factor is applied at the group (whole-plot) level and another at the individual (sub-plot) level. This design is particularly useful when some factors are harder or more expensive to randomize than others.

Example: Farming Experiment
A farm is testing two irrigation methods (Drip vs. Sprinkler) and two soil types (Clay vs. Sand) on crop yield. Since irrigation systems are installed at the farm level and are difficult to change, they are treated as the whole-plot factor. However, different soil types exist within each farm and can be tested more easily, making them the sub-plot factor.

Mathematical Model
The statistical model for a **Split-Plot Design** is:

$$Y_{ijk} = \mu + \alpha_i + B_k + \beta_j + (\alpha\beta)_{ij} + \epsilon_{ijk}$$

where:

- Y_{ijk} is the response (e.g., crop yield).
- μ is the overall mean.
- α_i is the **whole-plot factor** (Irrigation method).

- B_k is the **random block effect** (Farm-level variation).
- β_j is the **sub-plot factor** (Soil type).
- $(\alpha\beta)_{ij}$ is the interaction effect between **Irrigation** and **Soil type**.
- $\epsilon_{ijk} \sim N(0, \sigma^2)$ represents the **random error term**.

The key feature of the Split-Plot Design is that the whole-plot factor (α_i) is tested against the farm-level variance (B_k), while the sub-plot factor (β_j) is tested against individual variance (ϵ_{ijk}).

We model the **Split-Plot Design** using a **Mixed Effects Model**, treating **Farm** as a **random effect** to account for variation at the whole-plot level.

```
set.seed(123)

# Simulated dataset for a split-plot experiment
data <- data.frame(
  Farm = rep(1:6, each = 4), # 6 farms (whole plots)

  # Whole-plot factor
  Irrigation = rep(c("Drip", "Sprinkler"), each = 12),
  Soil = rep(c("Clay", "Sand"), times = 12), # Sub-plot factor

  # Response variable
  Yield = c(rnorm(12, mean=30, sd=5), rnorm(12, mean=35,
  ↪  sd=5))
)

# Load mixed-effects model library
library(lme4)

# Mixed-effects model: Whole-plot factor (Irrigation) as a
↪  random effect
model_split <- lmer(Yield ~ Irrigation * Soil + (1 | Farm),
↪  data = data)

# Summary of the model
summary(model_split)
#> Linear mixed model fit by REML ['lmerMod']
#> Formula: Yield ~ Irrigation * Soil + (1 | Farm)
#>    Data: data
#>
#> REML criterion at convergence: 128.1
#>
#> Scaled residuals:
#>      Min       1Q   Median       3Q      Max
```

```
#> -1.72562 -0.57572 -0.09767  0.60248  2.04346
#>
#> Random effects:
#>  Groups   Name        Variance Std.Dev.
#>  Farm     (Intercept)  0.00    0.000
#>  Residual             24.79    4.979
#> Number of obs: 24, groups:  Farm, 6
#>
#> Fixed effects:
#>                            Estimate Std. Error t value
#> (Intercept)                  31.771      2.033  15.629
#> IrrigationSprinkler           2.354      2.875   0.819
#> SoilSand                     -1.601      2.875  -0.557
#> IrrigationSprinkler:SoilSand  1.235      4.066   0.304
#>
#> Correlation of Fixed Effects:
#>             (Intr) IrrgtS SolSnd
#> IrrgtnSprnk -0.707
#> SoilSand    -0.707  0.500
#> IrrgtnSp:SS  0.500 -0.707 -0.707
#> optimizer (nloptwrap) convergence code: 0 (OK)
#> boundary (singular) fit: see help('isSingular')
```

In this model:

- Irrigation (Whole-plot factor) is tested against Farm-level variance.
- Soil type (Sub-plot factor) is tested against Residual variance.
- The interaction between Irrigation × Soil is also evaluated.

This hierarchical structure accounts for the fact that farms are not independent, improving the precision of our estimates.

3.4.6 Latin Square Design

When two potential confounding factors exist, Latin Square Designs provide a structured way to control for these variables. This design is common in scheduling, manufacturing, and supply chain experiments.

Example: Assembly Line Experiment
A manufacturer wants to test three assembly methods (A, B, C) while controlling for work shifts and workstations. Since both shifts and workstations may influence production time, a Latin Square Design ensures that each method is tested once per shift and once per workstation.

Mathematical Model

A Latin Square Design ensures that each treatment appears exactly once in each row and column:

$$Y_{ijk} = \mu + \alpha_i + \beta_j + \gamma_k + \epsilon_{ijk}$$

where:

- Y_{ijk} is the outcome (e.g., assembly time).
- μ is the overall mean.
- α_i is the **treatment effect** (Assembly Method).
- β_j is the **row effect** (Work Shift).
- γ_k is the **column effect** (Workstation).
- ϵ_{ijk} is the random error term.

This ensures that each treatment is **equally balanced across both confounding factors**.

We implement a **Latin Square Design** by treating **Assembly Method** as the primary factor, while controlling for **Shifts** and **Workstations**.

```
set.seed(123)

# Define the Latin Square layout
latin_square <- data.frame(
  Shift = rep(1:3, each = 3), # Rows
  Workstation = rep(1:3, times = 3), # Columns

  # Treatments assigned in a balanced way
  Method = c("A", "B", "C", "C", "A", "B", "B", "C", "A"),
  Time = c(rnorm(3, mean = 30, sd = 3),
           rnorm(3, mean = 28, sd = 3),
           rnorm(3, mean = 32, sd = 3)) # Assembly time
)

# ANOVA for Latin Square Design
anova_latin <-
  aov(Time ~ factor(Shift) + factor(Workstation) +
  ↪  factor(Method),
      data = latin_square)
summary(anova_latin)
#>                     Df Sum Sq Mean Sq F value Pr(>F)
#> factor(Shift)        2  1.148   0.574   0.079  0.927
#> factor(Workstation)  2 24.256  12.128   1.659  0.376
#> factor(Method)       2 14.086   7.043   0.964  0.509
#> Residuals            2 14.619   7.310
```

- If the p-value for "Method" is significant, we conclude that different assembly methods impact production time.
- If "Shift" or "Workstation" is significant, it indicates systematic differences across these variables.

3.5 Advanced Experimental Designs

3.5.1 Semi-Random Experiments

In **semi-random experiments**, participants are **not fully randomized** into treatment and control groups. Instead, a **structured randomization process** ensures fairness while allowing some level of causal inference.

3.5.1.1 Example: Loan Assignment Fairness

A bank wants to evaluate a new loan approval policy while ensuring that the experiment does not unfairly exclude specific demographics.
To maintain fairness:

1. Applicants are first stratified based on income and credit history.
2. Within each stratum, a random subset is assigned to receive the new policy, while others continue under the old policy.

```
set.seed(123)

# Create stratified groups
data <- data.frame(
  income_group = rep(c("Low", "Medium", "High"), each = 10),

  # Stratified randomization
  treatment = sample(rep(c("New Policy", "Old Policy"), each =
  ↪  15))
)

# Display the stratification results
table(data$income_group, data$treatment)
#>
#>          New Policy Old Policy
#>   High            6          4
#>   Low             6          4
#>   Medium          3          7
```

This approach ensures that each income group is fairly represented in both treatment and control conditions.

3.5.1.2 Case Study: Chicago Open Enrollment Program

A well-known example of semi-random assignment is the **Chicago Open Enrollment Program** (Cullen et al., 2005), where students could apply to choice schools.

However, since many schools were oversubscribed (i.e., demand exceeded supply), they used a random lottery system to allocate spots.

Thus, while **enrollment itself was not fully random**, the **lottery outcomes were random**, allowing researchers to estimate the Intent-to-Treat effect while acknowledging that not all students who won the lottery actually enrolled.

This situation presents a classic case where:

- **School choice is not random**: Families self-select into applying to certain schools.
- **Lottery outcomes are random**: Among those who apply, winning or losing the lottery is as good as a **random assignment**.

Let:

- $Enroll_{ij} = 1$ if student i enrolls in school j, and 0 otherwise.
- $Win_{ij} = 1$ if student i **wins the lottery**, and 0 otherwise.
- $Apply_{ij} = 1$ if student i **applies** to school j.

We define:

$$\delta_j = E[Y_i | Enroll_{ij} = 1, Apply_{ij} = 1] - E[Y_i | Enroll_{ij} = 0, Apply_{ij} = 1]$$

$$\theta_j = E[Y_i | Win_{ij} = 1, Apply_{ij} = 1] - E[Y_i | Win_{ij} = 0, Apply_{ij} = 1]$$

where:

- δ_j is the **treatment effect** (impact of actual school enrollment).
- θ_j is the intent-to-treat effect (impact of winning the lottery).

Since **not all winners enroll**, we know that:

$$\delta_j \neq \theta_j$$

Thus, we can only estimate θ_j directly and need an additional method to recover δ_j.

This distinction is crucial because simply comparing lottery winners and losers does not measure the true effect of enrollment—only the effect of being given the opportunity to enroll.

To estimate the treatment effect (δ_j), we use an Instrumental Variable approach:

$$\delta_j = \frac{E[Y_i|W_{ij} = 1, A_{ij} = 1] - E[Y_i|W_{ij} = 0, A_{ij} = 1]}{P(Enroll_{ij} = 1|W_{ij} = 1, A_{ij} = 1) - P(Enroll_{ij} = 1|W_{ij} = 0, A_{ij} = 1)}$$

where:

- $P(Enroll_{ij} = 1|W_{ij} = 1, A_{ij} = 1)$ = probability of enrolling **if winning the lottery**.
- $P(Enroll_{ij} = 1|W_{ij} = 0, A_{ij} = 1)$ = probability of enrolling **if losing the lottery**.

This adjustment accounts for the fact that some lottery winners do not enroll, and thus the observed effect (θ_j) underestimates the true treatment effect (δ_j).

This Instrumental Variable approach corrects for **selection bias** by leveraging **randomized lottery assignment**.

Numerical Example
Assume **10 students win** the lottery and **10 students lose**.

For Winners: (Table 3.1)
For Losers: (Table 3.2)

Computing Intent-to-Treat Effect

We compute the expected outcome for those who won and lost the lottery.

$$E[Y_i|W_{ij} = 1, A_{ij} = 1] = \frac{1(1.2) + 2(1) + 7(-0.1)}{10} = 0.25$$

$$E[Y_i|W_{ij} = 0, A_{ij} = 1] = \frac{1(1.2) + 2(0) + 7(-0.1)}{10} = 0.05$$

Thus, the Intent-to-Treat Effect is:

$$\text{Intent-to-Treat Effect} = 0.25 - 0.05 = 0.2$$

Table 3.1 Selection and treatment effects by compliance type winners

Type	Count	Selection effect	Treatment effect	Total effect
Always takers	1	+ 0.2	+ 1	+ 1.2
Compliers	2	0	+ 1	+ 1
Never takers	7	− 0.1	0	− 0.1

Table 3.2 Selection and treatment effects by compliance type losers

Type	Count	Selection effect	Treatment effect	Total effect
Always takers	1	+ 0.2	+ 1	+ 1.2
Compliers	2	0	0	0
Never takers	7	− 0.1	0	− 0.1

Now, we calculate the probability of enrollment for lottery winners and losers:

$$P(Enroll_{ij} = 1 | W_{ij} = 1, A_{ij} = 1) = \frac{1+2}{10} = 0.3$$

$$P(Enroll_{ij} = 1 | W_{ij} = 0, A_{ij} = 1) = \frac{1}{10} = 0.1$$

Using the formula for treatment effect (δ):

$$\text{Treatment Effect} = \frac{0.2}{0.3 - 0.1} = 1$$

This confirms that the true treatment effect is 1 unit.

To account for additional factors (X_i), we extend the model as follows:

$$Y_{ia} = \delta W_{ia} + \lambda L_{ia} + X_i\theta + u_{ia}$$

where:

- δ = Intent-to-Treat effect
- λ = True treatment effect
- W = Whether a student wins the lottery
- L = Whether a student enrolls in the school
- $X_i\theta$ = Control variables (i.e., reweighting of lottery), but would not affect treatment effect $E(\delta)$

Since choosing to apply to a lottery is not random, we must consider the following:

$$E(\lambda) \neq E(\lambda_1)$$

This demonstrates why lottery-based assignment is a useful but imperfect tool for causal inference—winning the lottery is random, but who applies is not.

3.5.2 *Re-randomization*

In standard randomization, baseline covariates are only balanced on average, meaning imbalance can still occur due to random chance. Re-randomization eliminates bad randomizations by checking balance before the experiment begins (Morgan and Rubin, 2012).

Key Motivations for Re-randomization

- **Randomization does not guarantee balance**:
  - Example: For **10 covariates**, the probability of at least one imbalance at $\alpha = 0.05$ is:

$$1 - (1 - 0.05)^{10} = 0.40 = 40\%$$

  - This means a high chance of some imbalance across treatment groups.
- **Re-randomization increases precision**: If covariates correlate with the outcome, improving covariate balance improves treatment effect estimation.
- **Accounting for re-randomization in inference**: Since re-randomization filters out bad assignments, it is equivalent to increasing the sample size and must be considered when computing standard errors.
- **Alternative balancing techniques**:
  - Stratified randomization (Johansson and Schultzberg, 2022)
  - Matched randomization (Greevy et al., 2004; Kapelner and Krieger, 2014)
  - Minimization (Pocock and Simon, 1975)

Example: Balancing Experimental Groups

An online retailer is testing two website designs (A and B) but wants to ensure that key customer demographics (e.g., age) are balanced across treatment groups.

We define a balance criterion to check if the mean age difference between groups is acceptable before proceeding.

```
set.seed(123)

# Define balance criterion: Ensure mean age difference < 1
↪  year
balance_criterion <- function(data) {
  abs(mean(data$age[data$group == "A"]) -
   ↪  mean(data$age[data$group == "B"])) < 1
}

# Generate randomized groups, repeat until balance criterion
↪  is met
repeat {
  data <- data.frame(
    age = rnorm(100, mean = 35, sd = 10),
    group = sample(c("A", "B"), 100, replace = TRUE)
  )
  if (balance_criterion(data)) break
}
```

```r
# Check final group means
tapply(data$age, data$group, mean)
#>        A        B
#> 35.91079 35.25483
```

3.5.2.1 Re-randomization Criterion

Re-randomization is based on a function of the covariate matrix (**X**) and treatment assignments (**W**).

$$W_i = \begin{cases} 1, & \text{if treated} \\ 0, & \text{if control} \end{cases}$$

A common approach is to use Mahalanobis Distance to measure covariate balance between treatment and control groups:

$$\begin{aligned} M &= (\bar{\mathbf{X}}_T - \bar{\mathbf{X}}_C)' \text{cov}(\bar{\mathbf{X}}_T - \bar{\mathbf{X}}_C)^{-1} (\bar{\mathbf{X}}_T - \bar{\mathbf{X}}_C) \\ &= (\frac{1}{n_T} + \frac{1}{n_C})^{-1} (\bar{\mathbf{X}}_T - \bar{\mathbf{X}}_C)' \text{cov}(\mathbf{X})^{-1} (\bar{\mathbf{X}}_T - \bar{\mathbf{X}}_C) \end{aligned}$$

where:

- $\bar{\mathbf{X}}_T$ and $\bar{\mathbf{X}}_C$ are the mean covariate values for treatment and control groups.
- cov(**X**) is the covariance matrix of the covariates.
- n_T and n_C are the sample sizes for treatment and control groups.

If the sample size is large and randomization is pure, then M follows a chi-squared distribution:

$$M \sim \chi^2_k$$

where k is the number of covariates to be balanced.

Choosing the Re-randomization Threshold ($M > a$
Define p_a as the probability of accepting a randomization:

- Smaller $p_a \Rightarrow$ Stronger balance, but longer computation time
- Larger $p_a \Rightarrow$ Faster randomization, but weaker balance

A rule of thumb is to re-randomize whenever:

$$M > a$$

where a is chosen based on acceptable balance thresholds.

We apply Mahalanobis Distance as a balance criterion to ensure that treatment and control groups are well-matched before proceeding with the experiment.

```
set.seed(123)
library(MASS)

# Generate a dataset with two covariates
n <- 100
X <- mvrnorm(n, mu = c(0, 0), Sigma = matrix(c(1, 0.5, 0.5,
↪ 1), 2, 2))
colnames(X) <- c("Covariate1", "Covariate2")

# Balance function using Mahalanobis Distance
balance_criterion <- function(X, group) {
  X_treat <- X[group == 1, ]
  X_control <- X[group == 0, ]

  mean_diff <- colMeans(X_treat) - colMeans(X_control)
  cov_inv <- solve(cov(X))
  M <- t(mean_diff) %*% cov_inv %*% mean_diff


  # Acceptable threshold
  # (chi-squared critical value for k = 2, alpha = 0.05)
  return(M < 3.84)
}

# Repeat randomization until balance is met
repeat {
  group <- sample(c(0, 1), n, replace = TRUE)
  if (balance_criterion(X, group)) break
}

# Display final balance check
table(group)
#> group
#>  0  1
#> 50 50
colMeans(X[group == 1, ])  # Treatment group means
#> Covariate1 Covariate2
#>  0.2469635  0.1918521
colMeans(X[group == 0, ])  # Control group means
#>  Covariate1  Covariate2
#>  0.01717088 -0.14281124
```

3.5.3 Two-Stage Randomized Experiments

A **two-stage randomized experiment** involves **sequential interventions**, where treatment assignments depend on earlier responses. This design is widely used in:

- **Adaptive Learning**: Adjusting educational content based on student progress
- **Personalized Advertising**: Targeting follow-up ads based on engagement
- **Medical Trials**: Adapting treatments based on patient response

By introducing **a second randomization stage**, researchers can evaluate:

1. **The effect of initial treatments**
2. **The effect of follow-up treatments**
3. **Potential interactions between the two stages**

3.5.3.1 Example: Personalized Advertising Experiment

A company tests two initial ad campaigns (Ad A vs. Ad B). After observing customer engagement, they apply a second-stage intervention (e.g., Discount vs. No Discount).

The two-stage experiment can be modeled as:

$$Y_{ijk} = \mu + \alpha_i + \beta_{j(i)} + \epsilon_{ijk}$$

where:

- μ = Overall mean outcome (e.g., conversion rate)
- α_i = Effect of first-stage intervention (i = Ad A or Ad B)
- $\beta_{j(i)}$ = Effect of second-stage intervention, nested within first-stage groups
- ϵ_{ijk} = Random error term

The nested structure ensures that the second-stage treatment ($\beta_{j(i)}$) is assigned within each first-stage treatment group.

```
set.seed(123)

# Generate first-stage randomization (Initial Ad)
data <- data.frame(
  stage1 = sample(c("Ad A", "Ad B"), 100, replace = TRUE),
  stage2 = rep(NA, 100)  # Placeholder for second-stage
   ↪  randomization
)

# Second-stage assignment based on first-stage response
data$stage2[data$stage1 == "Ad A"] <-
    sample(c("Discount", "No Discount"),
```

```
              sum(data$stage1 == "Ad A"),
              replace = TRUE)
data$stage2[data$stage1 == "Ad B"] <-
    sample(c("Discount", "No Discount"),
              sum(data$stage1 == "Ad B"),
              replace = TRUE)

# Display final randomization
table(data$stage1, data$stage2)
#>
#>          Discount No Discount
#>    Ad A        24          33
#>    Ad B        22          21
```

This structure ensures:

1. Fair assignment of initial ads
2. Adaptive targeting in the second stage based on user engagement

3.5.4 Two-Stage Randomized Experiments with Interference and Noncompliance

In real-world experiments, **interference and noncompliance** complicate analysis (Imai et al., 2021):

- **Interference**: When treatment effects "spill over" from one group to another (e.g., social influence in marketing)
- **Noncompliance**: When participants do not adhere to their assigned treatment (e.g., a customer ignoring an ad)

To handle noncompliance, we define:

- Z_{ik} = Assigned treatment (e.g., Ad A or Ad B)
- D_{ik} = Actual treatment received (e.g., whether the user actually saw the ad)
- Y_{ik} = Outcome (e.g., purchase)

A **two-stage Instrumental Variable model** adjusts for noncompliance:

$$D_{ik} = \gamma_0 + \gamma_1 Z_{ik} + v_{ik}$$

$$Y_{ik} = \beta_0 + \beta_1 D_{ik} + \epsilon_{ik}$$

where:

- γ_1 measures the effect of assignment on actual treatment received.
- β_1 estimates the treatment effect, adjusting for noncompliance.

If individuals influence each other's outcomes, traditional randomization is biased. Solutions include:

1. **Cluster Randomization**: Assigning treatments at the group level (e.g., entire social circles receive the same ad)
2. **Partial Interference Models**: Assume interference only occurs within predefined groups

```
set.seed(123)
library(ivreg)  # Load Instrumental Variable Regression
↪  Package

# Generate data for first-stage treatment assignment
n <- 500
data <- data.frame(
  Z = sample(c(0, 1), n, replace = TRUE),  # Initial
  ↪  assignment (randomized)
  D = NA,  # Actual treatment received (affected by
  ↪  compliance)
  Y = NA   # Outcome variable (e.g., purchase)
)

# Introduce noncompliance: 80% compliance rate
data$D <- ifelse(runif(n) < 0.8, data$Z, 1 - data$Z)

# Generate outcome variable (Y) with true treatment effect
# True effect of D on Y is 3
data$Y <- 5 + 3 * data$D + rnorm(n, mean = 0, sd = 2)

# Estimate Two-Stage Least Squares (2SLS)
iv_model <- ivreg(Y ~ D | Z, data = data)
summary(iv_model)
#>
#> Call:
#> ivreg(formula = Y ~ D | Z, data = data)
#>
#> Residuals:
#>     Min      1Q Median      3Q     Max
#> -5.497 -1.344  0.018  1.303  5.493
#>
#> Coefficients:
#>             Estimate Std. Error t value Pr(>|t|)
#> (Intercept)   5.1285     0.1861  27.557   <2e-16 ***
#> D             2.7487     0.3072   8.949   <2e-16 ***
#>
```

```
#> Diagnostic tests:
#>                     df1 df2 statistic p-value
#> Weak instruments      1 498    263.54  <2e-16 ***
#> Wu-Hausman            1 497      0.19   0.663
#> Sargan                0  NA        NA      NA
#> ---
#> Signif. codes:  0 '***' 0.001 '**' 0.01 '*' 0.05 '.' 0.1 ' ' 1
#>
#> Residual standard error: 2.017 on 498 degrees of freedom
#> Multiple R-Squared: 0.2997,  Adjusted R-squared: 0.2983
#> Wald test: 80.08 on 1 and 498 DF,  p-value: < 2.2e-16
```

- The first-stage regression estimates how strongly Z affects D (compliance).
- The second-stage regression estimates the true causal effect of D on Y.

If interference is present, the standard IV method may be biased. Researchers should explore network-based randomization or spatial models.

3.6 Emerging Research

Recent research highlights significant challenges in measuring the causal effects of ad content in online advertising experiments. Digital advertising platforms employ algorithmic targeting and dynamic ad delivery mechanisms, which can introduce systematic biases in A/B testing (Braun and Schwartz, 2025). Key concerns include:

- **Nonrandom Exposure**: Online advertising platforms optimize ad delivery dynamically, meaning users are not randomly assigned to different ad variants. Instead, platforms use proprietary algorithms to serve ads to different, often highly heterogeneous, user groups.
- **Divergent Delivery**: The optimization process can lead to "divergent delivery," where differences in user engagement patterns and platform objectives result in noncomparable treatment groups. This confounds the true effect of ad content with algorithmic biases in exposure.
- **Bias in Measured Effects**: Algorithmic targeting, user heterogeneity, and data aggregation can distort both the **magnitude** and **direction** of estimated ad effects. This means traditional A/B test results may not accurately reflect the true impact of ad creatives.
- **Limited Transparency**: Platforms have little incentive to assist advertisers or researchers in disentangling ad content effects from proprietary targeting mechanisms. As a result, experimenters must design careful identification strategies to mitigate these biases.

3.6.1 Covariate Balancing in Online A/B Testing: The Pigeonhole Design

Zhao and Zhou (2025) address the challenge of covariate balancing in online A/B testing when experimental subjects arrive sequentially. Traditional experimental designs struggle to maintain balance in real-time settings where treatment assignments must be made immediately. This work introduces the **online blocking problem**, in which subjects with heterogeneous covariates must be assigned to treatment or control groups dynamically, with the goal of minimizing **total discrepancy**—quantified as the minimum weight perfect matching between groups.

To improve covariate balance, the authors propose the **pigeonhole design**, a randomized experimental design that operates as follows:

- The covariate space is partitioned into smaller subspaces called "pigeonholes."
- Within each pigeonhole, the design balances the number of treated and control subjects, reducing systematic imbalances.

Theoretical analysis demonstrates the effectiveness of this design in reducing discrepancy compared to existing approaches. Specifically, the pigeonhole design is benchmarked against:

- **Matched-pair design**
- Completely Randomized Design

The results highlight scenarios where the pigeonhole design outperforms these alternatives in maintaining covariate balance.

Using Yahoo! data, the authors validate the pigeonhole design in practice, demonstrating a 10.2% reduction in variance when estimating the Average Treatment Effect. This improvement underscores the design's practical utility in reducing noise and improving the precision of experimental results.

The findings of Zhao and Zhou (2025) provide valuable insights for practitioners conducting online A/B tests:

- Ensuring real-time covariate balance improves the reliability of causal estimates.
- The pigeonhole design offers a structured yet flexible approach to balancing covariates dynamically.
- By reducing variance in treatment effect estimation, it enhances statistical efficiency in large-scale online experiments.

This study contributes to the broader discussion of randomized experimental design in dynamic settings, providing a compelling alternative to traditional approaches.

3.6.2 Handling Zero-Valued Outcomes

When analyzing treatment effects, a common issue arises when the outcome variable includes zero values.

For example, in business applications, a marketing intervention may have no effect on some customers (resulting in zero sales). If we were to apply a log transformation to the outcome variable, this would be problematic because:

- log(0) is undefined.
- Log transformation is sensitive to outcome units, making interpretation difficult (Chen and Roth, 2024).

Instead of applying a log transformation, we should use methods that are robust to zero values:

1. **Percentage Changes in the Average**
   - By using Poisson Quasi-Maximum Likelihood Estimation (QMLE), we can interpret coefficients as percentage changes in the mean outcome of the treatment group relative to the control group.
2. **Extensive vs. Intensive Margins**
   - This approach distinguishes between:
     - **Extensive margin**: The likelihood of moving from zero to a positive outcome (e.g., increasing the probability of making a sale)
     - **Intensive margin**: The increase in the outcome given that it is already positive (e.g., increasing the sales amount)
   - To estimate the **intensive-margin bounds**, we could use Lee (2009), assuming that treatment has a **monotonic effect** on the outcome.

We first generate a dataset to simulate a scenario where an outcome variable (e.g., sales, website clicks) has many zeros, and treatment affects both the extensive and intensive margins (Fig. 3.1).

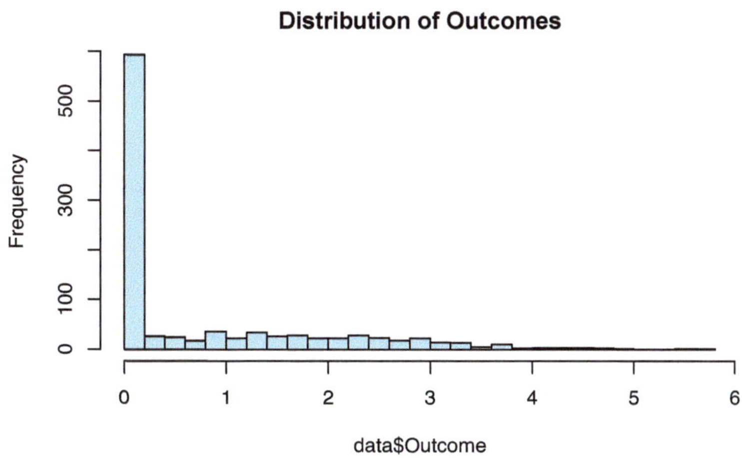

Fig. 3.1 Distribution of outcomes

```
set.seed(123) # For reproducibility
library(tidyverse)

n <- 1000 # Number of observations
p_treatment <- 0.5 # Probability of being treated

# Step 1: Generate the treatment variable D
D <- rbinom(n, 1, p_treatment)

# Step 2: Generate potential outcomes
# Untreated potential outcome (mostly zeroes)
Y0 <- rnorm(n, mean = 0, sd = 1) * (runif(n) < 0.3)

# Treated potential outcome (affecting both extensive and
↪  intensive margins)
Y1 <- Y0 + rnorm(n, mean = 2, sd = 1) * (runif(n) < 0.7)

# Step 3: Combine effects based on treatment assignment
Y_observed <- (1 - D) * Y0 + D * Y1

# Ensure non-negative outcomes (modeling real-world
↪  situations)
Y_observed[Y_observed < 0] <- 0

# Create a dataset with an additional control variable
data <- data.frame(
    ID = 1:n,
    Treatment = D,
    Outcome = Y_observed,
    X = rnorm(n) # Control variable
) |>
    dplyr::mutate(positive = Outcome > 0) # Indicator for
↪  extensive margin

# View first few rows
head(data)
#>   ID Treatment   Outcome          X positive
#> 1  1         0 0.0000000  1.4783345    FALSE
#> 2  2         1 2.2369379 -1.4067867     TRUE
#> 3  3         0 0.0000000 -1.8839721    FALSE
#> 4  4         1 3.2192276 -0.2773662     TRUE
#> 5  5         1 0.6649693  0.4304278     TRUE
#> 6  6         0 0.0000000 -0.1287867    FALSE
```

```r
# Plot distribution of outcomes
hist(
    data$Outcome,
    breaks = 30,
    main = "Distribution of Outcomes",
    col = "lightblue"
)
```

3.6.2.1 Estimating Percentage Changes in the Average

Since we cannot use a log transformation, we estimate percentage changes using Poisson QMLE, which is robust to zero-valued outcomes.

```r
library(fixest)

# Poisson Quasi-Maximum Likelihood Estimation (QMLE)
res_pois <- fepois(
    fml = Outcome ~ Treatment + X,
    data = data,
    vcov = "hetero"
)

# Display results
etable(res_pois)
#>                                      res_pois
#> Dependent Var.:                       Outcome
#>
#> Constant              -2.223*** (0.1440)
#> Treatment              2.579*** (0.1494)
#> X                        0.0235 (0.0406)
#> _______________  ___________________
#> S.E. type          Heteroskedas.-rob.
#> Observations                    1,000
#> Squared Cor.                  0.33857
#> Pseudo R2                     0.26145
#> BIC                           1,927.9
#> ---
#> Signif. codes: 0 '***' 0.001 '**' 0.01 '*' 0.05 '.' 0.1 ' ' 1
```

To interpret the results:

- The coefficient on `Treatment` represents the log-percentage change in the mean outcome of the treatment group relative to the control group.
- To compute the proportional effect, we exponentiate the coefficient:

```
# Compute the proportional effect of treatment
treatment_effect <- exp(coefficients(res_pois)["Treatment"]) - 1
treatment_effect
#> Treatment
#>  12.17757
```

Compute the standard error:

```
# Compute standard error
se_treatment <- exp(coefficients(res_pois)["Treatment"]) *
    sqrt(res_pois$cov.scaled["Treatment", "Treatment"])
se_treatment
#> Treatment
#>  1.968684
```

Thus, we conclude that the treatment effect increases the outcome by 1215% for the treated group compared to the control group.

3.6.2.2 Estimating Extensive Versus Intensive Margins

We now estimate treatment effects separately for the extensive margin (probability of a nonzero outcome) and the intensive margin (magnitude of effect among those with a nonzero outcome).

First, we use Lee Bounds to estimate the intensive-margin effect for individuals who always have a nonzero outcome.

```
library(causalverse)

res <- causalverse::lee_bounds(
    df = data,
    d = "Treatment",
    m = "positive",
    y = "Outcome",
    numdraws = 10
) |>
    causalverse::nice_tab(2)

print(res)
#>          term estimate std.error
#> 1 Lower bound    -0.22      0.09
#> 2 Upper bound     2.77      0.14
```

Since the confidence interval includes zero, we cannot conclude that the treatment has a significant intensive-margin effect.

3.6.2.3 Sensitivity Analysis: Varying the Effect for Compliers

To further investigate the intensive-margin effect, we consider how sensitive our results are to different assumptions about compliers.

We assume that the expected outcome for compliers is $100 \times c$ lower or higher than for always-takers:

$$E(Y(1)\|\text{Complier}) = (1 - c)E(Y(1)\|\text{Always-taker})$$

We compute Lee Bounds for different values of c.

```
set.seed(1)
c_values = c(0.1, 0.5, 0.7)

combined_res <- bind_rows(lapply(c_values, function(c) {
    res <- causalverse::lee_bounds(
        df = data,
        d = "Treatment",
        m = "positive",
        y = "Outcome",
        numdraws = 10,
        c_at_ratio = c
    )

    res$c_value <- as.character(c)
    return(res)
}))

combined_res |>
    dplyr::select(c_value, everything()) |>
    causalverse::nice_tab()
#>   c_value           term estimate std.error
#> 1     0.1 Point estimate     6.60      0.71
#> 2     0.5 Point estimate     2.54      0.13
#> 3     0.7 Point estimate     1.82      0.08
```

- If we assume $c = 0.1$, meaning compliers' outcomes are 10% of always-takers', then the intensive-margin effect is 6.6 units higher for always-takers.
- If $c = 0.5$, meaning compliers' outcomes are 50% of always-takers', then the intensive-margin effect is 2.54 units higher.

These results highlight how assumptions about compliers affect conclusions about the intensive margin.

When dealing with zero-valued outcomes, log transformations are not appropriate. Instead:

- **Poisson QMLE** provides a robust way to estimate percentage changes in the outcome.
- **Extensive vs. Intensive Margins** allow us to distinguish between:
  - The probability of a nonzero outcome (extensive margin)
  - The magnitude of change among those with a nonzero outcome (intensive margin)
- **Lee Bounds** provide a method to estimate the intensive-margin effect, though results can be sensitive to assumptions about always-takers and compliers.

Chapter 4
Sampling

Sampling is essential for making valid generalizations from data, and this chapter builds a thorough understanding of sampling theory and practice. The chapter covers population definitions, various sampling techniques, and their implications for inference. Probability sampling methods, including simple random sampling, stratified sampling, and cluster sampling, are compared with non-probability approaches such as convenience and snowball sampling. The importance of representativeness and bias is emphasized, particularly in business research where decisions often depend on customer or market samples. Concepts such as sampling distributions and standard errors are mathematically defined. Special attention is given to unequal probability sampling and its correction through weighting. The chapter concludes with strategies for balanced sampling and techniques for determining appropriate sample sizes for given levels of power and confidence.

4.1 Population and Sample

This is a refresher on terminology regarding sampling.

- **Population** (N): The complete set of all elements under study.
- **Sample** (n): A subset of the population selected for analysis.
- **Parameter**: A numerical measure that describes a characteristic of a population (e.g., population mean μ, population variance σ^2).
- **Statistic**: A numerical measure computed from a sample, used to estimate a population parameter (e.g., sample mean $\bar{x}$, sample variance s^2).

A well-chosen sample ensures that results generalize to the population, reducing **sampling bias**.

M. Nguyen, *Experimental Design*, https://doi.org/10.1007/978-3-032-01839-7_4

4.2 Probability Sampling

Probability sampling methods ensure that every element in the population has a known, nonzero probability of being selected. These methods are preferred in inferential statistics since they allow for the estimation of sampling error.

4.2.1 Simple Random Sampling

Simple Random Sampling (SRS) ensures that every element in the population has an equal chance of being selected. This can be done with replacement or without replacement, impacting whether an element can be chosen more than once.

Below is an example of drawing a simple random sample without replacement from a population of 100 elements:

```
set.seed(123)
population <- 1:100  # A population of 100 elements
sample_srs <- sample(population, size = 10, replace = FALSE)
sample_srs
#>  [1] 31 79 51 14 67 42 50 43 97 25
```

Advantages:

- Simple and easy to implement
- Ensures unbiased selection

Disadvantages:

- May not represent subgroups well, especially in heterogeneous populations
- Requires access to a complete list of the population

4.2.1.1 Using dplyr

The `sample_n()` function in `dplyr` allows for simple random sampling from a dataset.

```
library(dplyr)
iris_df <- iris
set.seed(1)
sample_n(iris_df, 5)  # Randomly selects 5 rows from the iris
↪  dataset
#>   Sepal.Length Sepal.Width Petal.Length Petal.Width
↪  Species
#> 1          5.8         2.7          4.1         1.0
↪  versicolor
```

```
#> 2          6.4          2.8          5.6          2.1
↪   virginica
#> 3          4.4          3.2          1.3          0.2
↪   setosa
#> 4          4.3          3.0          1.1          0.1
↪   setosa
#> 5          7.0          3.2          4.7          1.4
↪   versicolor
```

4.2.1.2 Using the sampling Package

The `sampling` package provides functions for random sampling with and without replacement.

```
library(sampling)
# Assign a unique ID to each row in the dataset
iris_df$id <- 1:nrow(iris_df)

# Simple random sampling without replacement
srs_sample <- srswor(10, length(iris_df$id))
# srs_sample

# Simple random sampling with replacement
srs_sample_wr <- srswr(10, length(iris_df$id))
# srs_sample_wr
```

4.2.1.3 Using the sampler Package

The `sampler` package provides additional functionality, such as oversampling to account for nonresponse.

```
library(sampler)
rsamp(albania, n = 260, over = 0.1, rep = FALSE)
```

4.2.1.4 Handling Missing Data in Sample Collection

To compare a sample with received (collected) data and identify missing elements:

```
alsample <- rsamp(df = albania, 544) # Initial sample
alreceived <- rsamp(df = alsample, 390) # Collected data
rmissing(sampdf = alsample, colldf = alreceived, col_name = qvKod)
```

4.2.2 *Stratified Sampling*

Stratified sampling involves dividing the population into distinct strata based on a characteristic (e.g., age, income level, region). A random sample is then drawn from each stratum, often in proportion to its size within the population. This method ensures that all subgroups are adequately represented, improving the precision of estimates.

The following example demonstrates stratified sampling where individuals belong to three different groups (A, B, C), and a random sample is drawn from each.

```
library(dplyr)

set.seed(123)
data <- data.frame(
  ID = 1:100,
  Group = sample(c("A", "B", "C"), 100, replace = TRUE)
)

# Stratified random sampling: selecting 10 elements per group
stratified_sample <- data %>%
  group_by(Group) %>%
  sample_n(size = 10)

# stratified_sample
```

Advantages:

- Ensures representation of all subgroups
- More precise estimates compared to Simple Random Sampling
- Reduces sampling error by accounting for population variability

Disadvantages:

- Requires prior knowledge of population strata
- More complex to implement than SRS

4.2.2.1 Using dplyr for Stratified Sampling

Sampling by Fixed Number of Rows
Here, we extract **five random observations** from each species in the `iris` dataset.

```
library(dplyr)

set.seed(123)
sample_iris <- iris %>%
  group_by(Species) %>%
  sample_n(5)  # Selects 5 samples per species

# sample_iris
```

Sampling by Fraction of Each Stratum
Instead of selecting a fixed number, we can sample 15% of each species.

```
set.seed(123)
sample_iris <- iris %>%
  group_by(Species) %>%
  sample_frac(size = 0.15)  # Selects 15% of each species

# sample_iris
```

4.2.2.2 Using the sampler Package

The `sampler` package allows stratified sampling with proportional allocation.

```
library(sampler)

# Stratified sample using proportional allocation without
↪  replacement
ssamp(df = albania, n = 360, strata = qarku, over = 0.1)
```

4.2.2.3 Handling Missing Data in Stratified Sampling

To identify the number of missing values by stratum between the initial sample and the collected data:

```r
alsample <- rsamp(df = albania, 544)  # Initial sample
alreceived <- rsamp(df = alsample, 390)  # Collected data

smissing(
  sampdf = alsample,
  colldf = alreceived,
  strata = qarku,    # Strata column
  col_name = qvKod  # Column for checking missing values
)
```

4.2.3 Systematic Sampling

Selects every *k*th element after a random starting point.

```r
k <- 10  # Select every 10th element
start <- sample(1:k, 1)  # Random start point
sample_systematic <- population[seq(start, length(population),
↪ by = k)]
```

Advantages:

- Simple to implement
- Ensures even coverage

Disadvantages:

- If data follows a pattern, bias may be introduced.

4.2.4 Cluster Sampling

Instead of selecting individuals, entire clusters (e.g., cities, schools) are randomly chosen, and all members of selected clusters are included.

```r
data$Cluster <- sample(1:10, 100, replace = TRUE)  # Assign 10
↪ clusters
chosen_clusters <- sample(1:10, size = 3)  # Select 3 clusters
cluster_sample <- filter(data, Cluster %in% chosen_clusters)
```

Advantages:

- Cost-effective when the population is large
- Useful when the population is naturally divided into groups

Disadvantages:

- Higher variability
- Risk of unrepresentative clusters

4.3 Non-Probability Sampling

These methods do not give all elements a known probability of selection. They are used in exploratory research but are not suitable for making formal statistical inferences.

4.3.1 Convenience Sampling

Selecting individuals who are easiest to reach (e.g., mall surveys).

Pros: Quick and inexpensive
Cons: High risk of bias, not generalizable

4.3.2 Quota Sampling

Similar to stratified sampling but nonrandom.

Pros: Ensures subgroup representation
Cons: Subject to selection bias

4.3.3 Snowball Sampling

Used for hard-to-reach populations (e.g., networking through referrals).

Pros: Useful when the population is unknown
Cons: High bias, dependency on initial subjects

4.4 Unequal Probability Sampling

Unequal probability sampling assigns different selection probabilities to elements in the population. This approach is often used when certain units are more important, have higher variability, or require higher precision in estimation.
Common methods for unequal probability sampling include:

- **Probability Proportional to Size (PPS)**: Selection probability is proportional to a given auxiliary variable (e.g., revenue, population size).
- **Poisson Sampling**: Independent selection of each unit with a given probability.
- **Systematic Sampling with Unequal Probabilities**: Uses a systematic approach while ensuring different probabilities.

The following functions from the `sampling` package implement various unequal probability sampling methods:

```
library(sampling)

# Different methods for unequal probability sampling
UPbrewer()          # Brewer's method
UPmaxentropy()      # Maximum entropy method
UPmidzuno()         # Midzuno's method
UPmidzunopi2()      # Midzuno's method with second-order
↪   inclusion probabilities
UPmultinomial()     # Multinomial method
UPpivotal()         # Pivotal method
UPrandompivotal()   # Randomized pivotal method
UPpoisson()         # Poisson sampling
UPsampford()        # Sampford's method
UPsystematic()      # Systematic sampling
UPrandomsystematic() # Randomized systematic sampling
UPsystematicpi2()   # Systematic sampling with second-order
↪   probabilities
UPtille()           # Tillé's method
UPtillepi2()        # Tillé's method with second-order
↪   inclusion probabilities
```

Each of these methods has specific use cases and theoretical justifications. For example:

- **Poisson sampling** allows flexible control over sample size but may lead to variable sample sizes.
- **Systematic sampling** is useful when population elements are arranged in a meaningful order.
- **Tillé's method** ensures better control over the sample's second-order inclusion probabilities.

4.5 Balanced Sampling

Balanced sampling ensures that the means of auxiliary variables in the sample match those in the population. This method improves estimation efficiency and reduces variability without introducing bias.

Balanced sampling differs from purposive selection because it still involves randomization, ensuring statistical validity.

The balancing equation is given by:

$$\sum_{k \in S} \frac{\mathbf{x}_k}{\pi_k} = \sum_{k \in U} \mathbf{x}_k$$

where:

- $\mathbf{x}_k$ is a vector of auxiliary variables (e.g., income, age, household size).
- π_k is the inclusion probability of unit k.
- S is the sample, and U is the population.

This ensures that the total weighted sum of auxiliary variables in the sample matches the total sum in the population.

4.5.1 *Cube Method for Balanced Sampling*

The **Cube Method** is a widely used approach for balanced sampling, consisting of two phases:

1. **Flight Phase**: Ensures initial balance on auxiliary variables.
2. **Landing Phase**: Adjusts the sample to meet constraints while keeping randomness.

```
library(sampling)

# Cube method functions
samplecube()         # Standard cube method
fastflightcube()     # Optimized flight phase
landingcube()        # Landing phase method
```

4.5.2 *Balanced Sampling with Stratification*

Stratification attempts to replicate the population structure in the sample by preserving the original multivariate histogram.

```
library(survey)
data("api")

# Stratified design with proportional allocation
srs_design <- svydesign(data = apistrat,
                        weights = ~pw,
                        fpc = ~fpc,
                        strata = ~stype,
                        id = ~1)
```

An additional method for balanced stratification is:

```
balancedstratification()
```

This method ensures that within each stratum, the sample retains the original proportions of auxiliary variables.

4.5.3 Balanced Sampling in Cluster Sampling

Cluster sampling involves selecting entire groups (clusters) instead of individual units. A balanced approach ensures that the sampled clusters preserve the overall distribution of auxiliary variables.

```
library(survey)
data("api")

# Cluster sampling design
srs_design <- svydesign(data = apiclus1,
                        weights = ~pw,
                        fpc = ~fpc,
                        id = ~dnum)
```

For explicitly balanced cluster sampling:

```
balancedcluster()
```

This method ensures that the cluster-level characteristics of the sample match those of the population.

4.5.4 Balanced Sampling in Two-Stage Sampling

Two-stage sampling first selects primary units (e.g., schools, cities) and then samples within them. A balanced approach ensures representative selection at both stages.

```
library(survey)
data("api")

# Two-stage sampling design
srs_design <- svydesign(data = apiclus2,
                        fpc = ~fpc1 + fpc2,
                        id = ~dnum + snum)
```

For explicitly balanced two-stage sampling:

```
balancedtwostage()
```

This method ensures that auxiliary variables remain balanced across both selection stages, reducing variability while maintaining randomness.

4.6 Sample Size Determination

The appropriate sample size depends on the margin of error, confidence level, and population variability. A commonly used formula for estimating the required sample size for a proportion is:

$$n = \frac{Z^2 p(1-p)}{E^2}$$

where:

- Z is the Z-score corresponding to the confidence level.
- p is the estimated proportion.
- E is the margin of error.

```
z <- qnorm(0.975)  # 95% confidence level
p <- 0.5  # Estimated proportion
E <- 0.05  # 5% margin of error
n <- (z^2 * p * (1 - p)) / (E^2)
ceiling(n)  # Round up to nearest integer
#> [1] 385
```

Chapter 5
Analysis of Variance

Analysis of Variance (ANOVA) forms a critical link between experimental design and statistical inference, and this chapter offers an in-depth look at its theoretical foundations and practical applications. Beginning with the Completely Randomized Design, the chapter introduces the partitioning of variance and the F-test, using intuitive visualizations and algebraic expressions. Advanced designs such as randomized blocks and nested designs are then developed, followed by nonparametric approaches that relax assumptions about normality and homogeneity. The mathematics behind ANOVA is derived in detail, including the expected mean squares and orthogonal decomposition of variance. Emphasis is placed on correct interpretation of p-values and effect sizes. The chapter concludes with guidance on sample size planning and the introduction of covariance models.

ANOVA shares its underlying mechanism with linear regression. However, ANOVA approaches the analysis from a different perspective, making it particularly useful for studying **qualitative variables** and **designed experiments**.

Key Terminology

- **Factor**: An explanatory or predictor variable studied in an experiment.
- **Treatment (Factor Level)**: A specific value or category of a factor applied to an experimental unit.
- **Experimental Unit**: The entity (e.g., person, animal, material) subjected to treatments and providing a response.
- **Single-Factor Experiment**: An experiment with only one explanatory variable.
- **Multifactor Experiment**: An experiment involving multiple explanatory variables.
- **Classification Factor**: A factor that is not controlled by the experimenter (common in observational studies).
- **Experimental Factor**: A factor that is directly assigned by the experimenter.

A well-designed experiment requires careful planning in the following areas:

- **Choice of Treatments**: Selecting factor levels to be tested.

M. Nguyen, *Experimental Design*, https://doi.org/10.1007/978-3-032-01839-7_5

- **Selection of Experimental Units**: Ensuring an appropriate sample.
- **Treatment Assignment**: Avoiding selection bias through proper randomization.
- **Measurement**: Minimizing measurement bias and considering blinding when necessary.

Advancements in Experimental Design

1. **Factorial Experiments**:
   - Investigate multiple factors simultaneously.
   - Allow for the study of interactions between factors.
2. **Replication**:
   - Repeating experiments increases statistical power.
   - Helps estimate mean squared error.
3. **Randomization**:
   - Introduced formally by R.A. Fisher in the early 1900s.
   - Ensures that treatment assignment is not systematically biased.
   - Helps eliminate confounding effects due to time, space, or other lurking variables.
4. **Local Control (Blocking/Stratification)**:
   - Reduces experimental error by controlling for known sources of variability.
   - Increases power by grouping similar experimental units together before randomizing treatments.

Randomization also helps eliminate correlations due to time and space.

5.1 Completely Randomized Design

A **Completely Randomized Design (CRD)** is the simplest type of experimental design, where experimental units are randomly assigned to treatments.

Consider a treatment factor A with $a \geq 2$ treatment levels. Each experimental unit is randomly assigned to one of these levels. The number of units in each group can be:

- **Balanced**: All groups have equal sample sizes n.
- **Unbalanced**: Groups have different sample sizes n_i (for $i = 1, ..., a$).

The total sample size is given by:

$$N = \sum_{i=1}^{a} n_i$$

Table 5.1 Treatment response table

Treatment	1	2	...	a
	Y_{11}	Y_{21}	...	Y_{a1}
	Y_{12}	...	...	...
	...	...	...	...
Sample mean	$\bar{Y_{1.}}$	$\bar{Y_{2.}}$	...	$\bar{Y_{a.}}$
Sample SD	s_1	s_2	...	s_a

Table 5.1 shows the treatment response table, where we can observe the treatment outcomes and their sample means and sample standard deviations. The number of possible assignments of units to treatments is:

$$k = \frac{N!}{n_1!n_2!\dots n_a!}$$

Each assignment has an equal probability of being selected: $1/k$. The response of each experimental unit is denoted as Y_{ij}, where:

- i indexes the treatment group.
- j indexes the individual unit within treatment i.

Where:

$$\bar{Y_{i.}} = \frac{1}{n_i} \sum_{j=1}^{n_i} Y_{ij}$$

$$s_i^2 = \frac{1}{n_i - 1} \sum_{j=1}^{n_i} (Y_{ij} - \bar{Y_{i.}})^2$$

The grand mean is:

$$\bar{Y_{..}} = \frac{1}{N} \sum_i \sum_j Y_{ij}$$

5.1.1 *Single-Factor Fixed Effects ANOVA*

Also known as **One-Way ANOVA** or **ANOVA Type I Model**.

The total variability in the response variable Y_{ij} can be decomposed as follows:

$$\begin{aligned} Y_{ij} - \bar{Y_{..}} &= Y_{ij} - \bar{Y_{..}} + \bar{Y_{i.}} - \bar{Y_{i.}} \\ &= (\bar{Y_{i.}} - \bar{Y_{..}}) + (Y_{ij} - \bar{Y_{i.}}) \end{aligned}$$

where:

- The first term represents **between-treatment variability** (deviation of treatment means from the grand mean).
- The second term represents **within-treatment variability** (deviation of observations from their treatment mean).

Thus, we partition the **total sum of squares (SSTO)** as:

$$\sum_i \sum_j (Y_{ij} - \bar{Y}_{..})^2 = \sum_i n_i(\bar{Y}_{i.} - \bar{Y}_{..})^2 + \sum_i \sum_j (Y_{ij} - \bar{Y}_{i.})^2$$

Or equivalently:

$$SSTO = SSTR + SSE$$

Where:

- **SSTO (Total SS)**: Total variability in the data.
- **SSTR (Treatment SS)**: Variability due to differences between treatment means.
- **SSE (Error SS)**: Variability within treatments (unexplained variance).

Degrees of freedom (df):

$$(N-1) = (a-1) + (N-a)$$

where we lose a degree of freedom for the total corrected SSTO because of the estimation of the mean ($\sum_i \sum_j (Y_{ij} - \bar{Y}_{..}) = 0$) and for the SSTR ($\sum_i n_i(\bar{Y}_{i.} - \bar{Y}_{..}) = 0$) (Table 5.2).

Mean squares:

$$MSTR = \frac{SSTR}{a-1}, \quad MSR = \frac{SSE}{N-a}$$

For a linear model interpretation of ANOVA, we have either

1. Cell Means Model
2. Treatment Effect (Factor Effects Model)

Table 5.2 ANOVA table

Source of variation	SS	df	MS
Between treatments	$\sum_i n_i(\bar{Y}_{i.} - \bar{Y}_{..})^2$	$a-1$	$SSTR/(a-1)$
Error (within treatments)	$\sum_i \sum_j (Y_{ij} - \bar{Y}_{i.})^2$	$N-a$	$SSE/(N-a)$
Total (corrected)	$\sum_i n_i(\bar{Y}_{i.} - \bar{Y}_{..})^2$	$N-1$	

5.1.1.1 Cell Means Model

The **cell means model** describes the response as:

$$Y_{ij} = \mu_i + \epsilon_{ij}$$

where:

- Y_{ij}: Response for unit j in treatment i.
- μ_i: Fixed population mean for treatment i.
- $\epsilon_{ij} \sim N(0, \sigma^2)$: Independent errors.
- $E(Y_{ij}) = \mu_i$, $\text{Var}(Y_{ij}) = \sigma^2$.

All observations are assumed to have **equal variance** across treatments.

Example: ANOVA with $a = 3$ Treatments

Consider a case with **three treatments** ($a = 3$), where each treatment has **two replicates** ($n_1 = n_2 = n_3 = 2$). The response vector can be expressed in matrix form as:

$$\begin{pmatrix} Y_{11} \\ Y_{12} \\ Y_{21} \\ Y_{22} \\ Y_{31} \\ Y_{32} \end{pmatrix} = \begin{pmatrix} 1\,0\,0 \\ 1\,0\,0 \\ 0\,1\,0 \\ 0\,1\,0 \\ 0\,0\,1 \\ 0\,0\,1 \end{pmatrix} \begin{pmatrix} \mu_1 \\ \mu_2 \\ \mu_3 \end{pmatrix} + \begin{pmatrix} \epsilon_{11} \\ \epsilon_{12} \\ \epsilon_{21} \\ \epsilon_{22} \\ \epsilon_{31} \\ \epsilon_{32} \end{pmatrix}$$

$$\mathbf{y} = \mathbf{X}\boldsymbol{\beta} + \boldsymbol{\epsilon}$$

where:

- $X_{k,ij} = 1$ if the k-th treatment is applied to unit (i, j).
- $X_{k,ij} = 0$ otherwise.

Note: There is **no intercept** term in this model.

The least squares estimator for β is given by:

$$\begin{aligned} \mathbf{b} = \begin{bmatrix} \mu_1 \\ \mu_2 \\ \mu_3 \end{bmatrix} &= (\mathbf{X}'\mathbf{X})^{-1}\mathbf{X}'\mathbf{y} \\ &= \begin{bmatrix} n_1 & 0 & 0 \\ 0 & n_2 & 0 \\ 0 & 0 & n_3 \end{bmatrix}^{-1} \begin{bmatrix} Y_1 \\ Y_2 \\ Y_3 \end{bmatrix} \\ &= \begin{bmatrix} \bar{Y_1} \\ \bar{Y_2} \\ \bar{Y_3} \end{bmatrix} \end{aligned} \tag{5.1}$$

Thus, the estimated treatment means are:

$$\hat{\mu}_i = \bar{Y}_i, \quad i = 1, 2, 3$$

This estimator $\mathbf{b} = [\bar{Y}_1, \bar{Y}_2, \bar{Y}_3]'$ is the **best linear unbiased estimator (BLUE)** for β (i.e., $E(\mathbf{b}) = \beta$)

Since $\mathbf{b} \sim N(\beta, \sigma^2(\mathbf{X}'\mathbf{X})^{-1})$, the variance of the estimated treatment means is:

$$var(\mathbf{b}) = \sigma^2(\mathbf{X}'\mathbf{X})^{-1} = \sigma^2 \begin{bmatrix} 1/n_1 & 0 & 0 \\ 0 & 1/n_2 & 0 \\ 0 & 0 & 1/n_3 \end{bmatrix}$$

Thus, the variance of each estimated treatment mean is:

$$var(b_i) = var(\hat{\mu}_i) = \frac{\sigma^2}{n_i}, \quad i = 1, 2, 3$$

The mean squared error (MSE) is given by:

$$\begin{aligned} MSE &= \frac{1}{N-a} \sum_{i=1}^{a} \sum_{j=1}^{n_i} (Y_{ij} - \overline{Y}_{i\cdot})^2 \\ &= \frac{1}{N-a} \sum_{i=1}^{a} \Big[(n_i - 1) \underbrace{\frac{1}{n_i - 1} \sum_{j=1}^{n_i} (Y_{ij} - \overline{Y}_{i\cdot})^2}_{= s_i^2} \Big] \\ &= \frac{1}{N-a} \sum_{i=1}^{a} (n_i - 1)\, s_i^2. \end{aligned}$$

where s_i^2 is the sample variance within the i-th treatment group.

Since $E(s_i^2) = \sigma^2$, we get:

$$E(MSE) = \frac{1}{N-a} \sum_i (n_i - 1)\sigma^2 = \sigma^2$$

Thus, MSE is an unbiased estimator of σ^2, regardless of whether the treatment means are equal.

The expected mean square for treatments (MSTR) is:

$$E(MSTR) = \sigma^2 + \frac{\sum_i n_i(\mu_i - \mu_.)^2}{a - 1}$$

where:

$$\mu_. = \frac{\sum_{i=1}^{a} n_i \mu_i}{\sum_{i=1}^{a} n_i}$$

If all treatment means are equal ($\mu_1 = \mu_2 = \cdots = \mu_a = \mu_.$), then:

$$E(MSTR) = \sigma^2$$

F-Test for Equality of Treatment Means
We test the null hypothesis:

$$H_0 : \mu_1 = \mu_2 = \cdots = \mu_a$$

against the alternative:

$$H_a : \text{at least one } \mu_i \text{ differs}$$

The test statistic is:

$$F = \frac{MSTR}{MSE}$$

- Large values of F suggest rejecting H_0 (since MSTR will be larger than MSE when H_a is true).
- Values of F near 1 suggest that we fail to reject H_0.

Since $MSTR$ and MSE are independent chi-square random variables scaled by their degrees of freedom, under H_0:

$$F \sim F_{(a-1,N-a)}$$

Decision Rule:

- If $F \leq F_{(a-1,N-a;1-\alpha)}$, fail to reject H_0.
- If $F \geq F_{(a-1,N-a;1-\alpha)}$, reject H_0.

If there are only two treatments ($a = 2$), the ANOVA F-test reduces to the two-sample t-test:

$$F = t^2$$

where:

$$t = \frac{\bar{Y_1} - \bar{Y_2}}{\sqrt{MSE\left(\frac{1}{n_1} + \frac{1}{n_2}\right)}}$$

5.1.1.2 Treatment Effects (Factor Effects)

Besides cell means model, we have another way to formalize one-way ANOVA:

$$Y_{ij} = \mu + \tau_i + \epsilon_{ij}$$

where:

- Y_{ij} is the j-th response for the i-th treatment.
- τ_i is the i-th treatment effect.
- μ is the constant component common to all observations.
- ϵ_{ij} are independent random errors, assumed to be normally distributed: $\epsilon_{ij} \sim N(0, \sigma^2)$.

For example, if we have $a = 3$ treatments and $n_1 = n_2 = n_3 = 2$ observations per treatment, the model representation is:

$$\begin{pmatrix} Y_{11} \\ Y_{12} \\ Y_{21} \\ Y_{22} \\ Y_{31} \\ Y_{32} \end{pmatrix} = \begin{pmatrix} 1\,1\,0\,0 \\ 1\,1\,0\,0 \\ 1\,0\,1\,0 \\ 1\,0\,1\,0 \\ 1\,0\,0\,1 \\ 1\,0\,0\,1 \end{pmatrix} \begin{pmatrix} \mu \\ \tau_1 \\ \tau_2 \\ \tau_3 \end{pmatrix} + \begin{pmatrix} \epsilon_{11} \\ \epsilon_{12} \\ \epsilon_{21} \\ \epsilon_{22} \\ \epsilon_{31} \\ \epsilon_{32} \end{pmatrix} \tag{5.2}$$

$$\mathbf{y} = \mathbf{X}\boldsymbol{\beta} + \boldsymbol{\epsilon}$$

However, the matrix:

$$\mathbf{X}'\mathbf{X} = \begin{pmatrix} \sum_i n_i & n_1 & n_2 & n_3 \\ n_1 & n_1 & 0 & 0 \\ n_2 & 0 & n_2 & 0 \\ n_3 & 0 & 0 & n_3 \end{pmatrix}$$

is singular, meaning $\mathbf{X}'\mathbf{X}$ is not invertible. This results in an infinite number of possible solutions for $\mathbf{b}$.

To resolve this, we impose restrictions on the parameters to ensure that $\mathbf{X}$ has full rank. Regardless of the restriction used, the expected value remains:

$$E(Y_{ij}) = \mu + \tau_i = \mu_i = \text{mean response for the } i\text{-th treatment}$$

5.1.1.2.1 Restriction on Sum of Treatment Effects

One common restriction is:

$$\sum_{i=1}^{a} \tau_i = 0$$

which implies that:

$$\mu = \frac{1}{a}\sum_{i=1}^{a}(\mu + \tau_i)$$

meaning that μ represents the grand mean (the overall mean response across treatments).

Each treatment effect can then be expressed as:

$$\begin{aligned}\tau_i &= \mu_i - \mu \\ &= \text{treatment mean} - \text{grand mean}\end{aligned}$$

Since $\sum_i \tau_i = 0$, we can solve for τ_a as:

$$\tau_a = -(\tau_1 + \tau_2 + \cdots + \tau_{a-1})$$

Thus, the mean for the a-th treatment is:

$$\mu_a = \mu + \tau_a = \mu - (\tau_1 + \tau_2 + \cdots + \tau_{a-1})$$

This reduces the number of parameters from $a+1$ to just a, meaning we estimate:

$$\mu, \tau_1, \tau_2, \ldots, \tau_{a-1}$$

Rewriting Equation (5.2):

$$\begin{pmatrix} Y_{11} \\ Y_{12} \\ Y_{21} \\ Y_{22} \\ Y_{31} \\ Y_{32} \end{pmatrix} = \begin{pmatrix} 1 & 1 & 0 \\ 1 & 1 & 0 \\ 1 & 0 & 1 \\ 1 & 0 & 1 \\ 1 & -1 & -1 \\ 1 & -1 & -1 \end{pmatrix} \begin{pmatrix} \mu \\ \tau_1 \\ \tau_2 \end{pmatrix} + \begin{pmatrix} \epsilon_{11} \\ \epsilon_{12} \\ \epsilon_{21} \\ \epsilon_{22} \\ \epsilon_{31} \\ \epsilon_{32} \end{pmatrix} \tag{5.3}$$

$$\mathbf{y} = \mathbf{X}\boldsymbol{\beta} + \boldsymbol{\epsilon}$$

where $\beta = [\mu, \tau_1, \tau_2]'$.

5.1.1.2.2 Restriction on the First τ

In **R**, the default parameterization in `lm()` for a one-way ANOVA model sets $\tau_1 = 0$. This effectively chooses the first treatment (or group) as a baseline or reference, making its treatment effect τ_1 equal to zero.

Consider the last example with three treatments, each having two observations, $n_1 = n_2 = n_3 = 2$. Under the restriction $\tau_1 = 0$, the treatment means can be expressed as:

$$\begin{aligned}
\mu_1 &= \mu + \tau_1 = \mu + 0 = \mu, \\
\mu_2 &= \mu + \tau_2, \\
\mu_3 &= \mu + \tau_3.
\end{aligned}$$

Hence, μ becomes the mean response for the **first** treatment.

We write the observations in vector form:

$$\begin{aligned}
\mathbf{y} = \begin{pmatrix} Y_{11} \\ Y_{12} \\ Y_{21} \\ Y_{22} \\ Y_{31} \\ Y_{32} \end{pmatrix} &= \underbrace{\begin{pmatrix} 1\,0\,0 \\ 1\,0\,0 \\ 1\,1\,0 \\ 1\,1\,0 \\ 1\,0\,1 \\ 1\,0\,1 \end{pmatrix}}_{\mathbf{X}} \begin{pmatrix} \mu \\ \tau_2 \\ \tau_3 \end{pmatrix} + \begin{pmatrix} \epsilon_{11} \\ \epsilon_{12} \\ \epsilon_{21} \\ \epsilon_{22} \\ \epsilon_{31} \\ \epsilon_{32} \end{pmatrix} \\
&= \mathbf{X}\boldsymbol{\beta} + \boldsymbol{\epsilon},
\end{aligned}$$

where

$$\beta = \begin{pmatrix} \mu \\ \tau_2 \\ \tau_3 \end{pmatrix}.$$

The OLS estimator is:

$$\mathbf{b} = \begin{pmatrix} \hat{\mu} \\ \hat{\tau_2} \\ \hat{\tau_3} \end{pmatrix} = (\mathbf{X}'\mathbf{X})^{-1}\mathbf{X}'\,\mathbf{y}.$$

In our specific case with equal sample sizes ($n_1 = n_2 = n_3 = 2$), the $(\mathbf{X}'\mathbf{X})^{-1}\mathbf{X}'\mathbf{y}$ calculation yields:

$$\mathbf{b} = \begin{bmatrix} \sum_i n_i & n_2 & n_3 \\ n_2 & n_2 & 0 \\ n_3 & 0 & n_3 \end{bmatrix}^{-1} \begin{bmatrix} Y_{..} \\ Y_{2.} \\ Y_{3.} \end{bmatrix}$$

$$= \begin{pmatrix} \bar{Y}_{1\cdot} \\ \bar{Y}_{2\cdot} - \bar{Y}_{1\cdot} \\ \bar{Y}_{3\cdot} - \bar{Y}_{1\cdot} \end{pmatrix}$$

where $\bar{Y}_{1\cdot}$, $\bar{Y}_{2\cdot}$, and $\bar{Y}_{3\cdot}$ are the sample means for treatments 1, 2, and 3, respectively.

Taking the expectation of $\mathbf{b}$ confirms:

$$E(\mathbf{b}) = \beta = \begin{pmatrix} \mu \\ \tau_2 \\ \tau_3 \end{pmatrix} = \begin{pmatrix} \mu_1 \\ \mu_2 - \mu_1 \\ \mu_3 - \mu_1 \end{pmatrix}.$$

Recall that:

$$\text{Var}(\mathbf{b}) = \sigma^2 (\mathbf{X}'\mathbf{X})^{-1}.$$

Hence,

$$\text{Var}(\hat{\mu}) = \text{Var}(\bar{Y}_{1\cdot}) = \frac{\sigma^2}{n_1},$$

$$\text{Var}(\hat{\tau}_2) = \text{Var}\big(\bar{Y}_{2\cdot} - \bar{Y}_{1\cdot}\big) = \frac{\sigma^2}{n_2} + \frac{\sigma^2}{n_1},$$

$$\text{Var}(\hat{\tau}_3) = \text{Var}\big(\bar{Y}_{3\cdot} - \bar{Y}_{1\cdot}\big) = \frac{\sigma^2}{n_3} + \frac{\sigma^2}{n_1}.$$

5.1.1.3 Equivalence of Parameterizations

Despite having different ways of writing the model, all three parameterizations yield the same ANOVA table:

1. Model 1: $Y_{ij} = \mu_i + \epsilon_{ij}$.
2. Model 2: $Y_{ij} = \mu + \tau_i + \epsilon_{ij}$, where $\sum_i \tau_i = 0$.
3. Model 3: $Y_{ij} = \mu + \tau_i + \epsilon_{ij}$, where $\tau_1 = 0$.

Table 5.3 Decomposition of variance in one-way analysis

Source of variation	SS	df	MS	F
Between treatments	$\sum_i n_i(\overline{Y}_{i\cdot} - \overline{Y}_{\cdot\cdot})^2 = \mathbf{Y}'(\mathbf{P} - \mathbf{P}_1)\mathbf{Y}$	$a-1$	$\frac{SSTR}{a-1}$	$\frac{MSTR}{MSE}$
Error (within treatments)	$\sum_i \sum_j (Y_{ij} - \overline{Y}_{i\cdot})^2 = \mathbf{e}'\mathbf{e}$	$N-a$	$\frac{SSE}{N-a}$	
Total (corrected)	$\sum_i n_i(\overline{Y}_{i\cdot} - \overline{Y}_{\cdot\cdot})^2 = \mathbf{Y}'\mathbf{Y} - \mathbf{Y}'\mathbf{P}_1\mathbf{Y}$	$N-1$		

All three lead to the same fitted values, because

$$\hat{\mathbf{Y}} = \mathbf{X}(\mathbf{X}'\mathbf{X})^{-1}\mathbf{X}'\mathbf{Y} = \mathbf{P}\mathbf{Y} = \mathbf{X}\mathbf{b}.$$

5.1.1.4 ANOVA Table

The generic form of the ANOVA table is (Table 5.3):
where $\mathbf{P}_1 = \frac{1}{n}\mathbf{J}$, $n = \sum_i n_i$, and $\mathbf{J}$ is the all-ones matrix.

The F-statistic has $(a-1, N-a)$ degrees of freedom and the numeric value is unchanged under any of the three parameterizations. The slight difference lies in how we state the null hypothesis:

$$H_0 : \mu_1 = \mu_2 = \cdots = \mu_a,$$

$$H_0 : \mu + \tau_1 = \mu + \tau_2 = \cdots = \mu + \tau_a,$$

$$H_0 : \tau_1 = \tau_2 = \cdots = \tau_a.$$

The F-test here serves as a preliminary analysis, to see if there is any difference at different factors. For more in-depth analysis, we consider different testing of treatment effects.

5.1.1.5 Testing of Treatment Effects

- Single Treatment Mean μ_i
- Differences Between Treatment Means
- Contrast Among Treatment Means
- Linear Combination of Treatment Means

5.1.1.5.1 Single Treatment Mean

For a single treatment group, the sample mean serves as an estimate of the population mean:

$$\hat{\mu_i} = \bar{Y}_{i.}$$

where:

- $E(\bar{Y}_{i.}) = \mu_i$, indicating unbiasedness.
- $var(\bar{Y}_{i.}) = \frac{\sigma^2}{n_i}$, estimated by $s^2(\bar{Y}_{i.}) = \frac{MSE}{n_i}$.

Since the standardized test statistic

$$T = \frac{\bar{Y}_{i.} - \mu_i}{s(\bar{Y}_{i.})}$$

follows a t-distribution with $N - a$ degrees of freedom (t_{N-a}), a $(1 - \alpha)100\%$ confidence interval for μ_i is:

$$\bar{Y}_{i.} \pm t_{1-\alpha/2;N-a} s(\bar{Y}_{i.})$$

To test whether μ_i is equal to some constant c, we set up the hypothesis:

$$H_0 : \mu_i = c$$

$$H_1 : \mu_i \neq c$$

The test statistic:

$$T = \frac{\bar{Y}_{i.} - c}{s(\bar{Y}_{i.})} \sim t_{N-a}$$

Under H_0, we reject H_0 at the α level if:

$$|T| > t_{1-\alpha/2;N-a}$$

5.1.1.5.2 Differences Between Treatment Means

The difference between two treatment means, also called a **pairwise comparison**, is given by:

$$D = \mu_i - \mu_{i'}$$

which is estimated by:

$$\hat{D} = \bar{Y}_{i.} - \bar{Y}_{i'.}$$

This estimate is unbiased since:

$$E(\hat{D}) = \mu_i - \mu_{i'}$$

Since $\bar{Y}_{i.}$ and $\bar{Y}_{i'.}$ are independent, the variance of $\hat{D}$ is:

$$var(\hat{D}) = var(\bar{Y}_{i.}) + var(\bar{Y}_{i'.}) = \sigma^2 \left(\frac{1}{n_i} + \frac{1}{n_{i'}}\right)$$

which is estimated by:

$$s^2(\hat{D}) = MSE \left(\frac{1}{n_i} + \frac{1}{n_{i'}}\right)$$

Using the same inference structure as the single treatment mean:

$$\frac{\hat{D} - D}{s(\hat{D})} \sim t_{N-a}$$

A $(1 - \alpha)100\%$ confidence interval for D is:

$$\hat{D} \pm t_{1-\alpha/2; N-a} s(\hat{D})$$

For hypothesis testing:

$$H_0 : \mu_i = \mu_{i'}$$
$$H_a : \mu_i \neq \mu_{i'}$$

we use the test statistic:

$$T = \frac{\hat{D}}{s(\hat{D})} \sim t_{N-a}$$

We reject H_0 at the α level if:

$$|T| > t_{1-\alpha/2; N-a}$$

5.1.1.5.3 Contrast Among Treatment Means

To generalize the comparison of two means, we introduce **contrasts**.

A **contrast** is a linear combination of treatment means:

$$L = \sum_{i=1}^{a} c_i \mu_i$$

where the coefficients c_i are nonrandom constants that satisfy the constraint:

$$\sum_{i=1}^{a} c_i = 0$$

This ensures that contrasts focus on relative comparisons rather than absolute magnitudes.

An unbiased estimator of L is given by:

$$\hat{L} = \sum_{i=1}^{a} c_i \bar{Y}_{i.}$$

Since expectation is a linear operator:

$$E(\hat{L}) = \sum_{i=1}^{a} c_i E(\bar{Y}_{i.}) = \sum_{i=1}^{a} c_i \mu_i = L$$

Thus, $\hat{L}$ is an unbiased estimator of L.

Since the sample means $\bar{Y}_{i.}$ are independent, the variance of $\hat{L}$ is:

$$\begin{aligned} var(\hat{L}) &= var\left(\sum_{i=1}^{a} c_i \bar{Y}_{i.}\right) \\ &= \sum_{i=1}^{a} c_i^2 var(\bar{Y}_{i.}) \\ &= \sum_{i=1}^{a} c_i^2 \frac{\sigma^2}{n_i} \\ &= \sigma^2 \sum_{i=1}^{a} \frac{c_i^2}{n_i} \end{aligned}$$

Since σ^2 is unknown, we estimate it using the mean squared error:

$$s^2(\hat{L}) = MSE \sum_{i=1}^{a} \frac{c_i^2}{n_i}$$

Since $\hat{L}$ is a linear combination of independent normal random variables, it follows a normal distribution:

$$\hat{L} \sim N\left(L, \sigma^2 \sum_{i=1}^{a} \frac{c_i^2}{n_i}\right)$$

Since $SSE/\sigma^2 \sim \chi^2_{N-a}$ and $MSE = SSE/(N-a)$, we use the t-distribution:

$$\frac{\hat{L} - L}{s(\hat{L})} \sim t_{N-a}$$

Thus, a $(1-\alpha)100\%$ confidence interval for L is:

$$\hat{L} \pm t_{1-\alpha/2;N-a} s(\hat{L})$$

To test whether a specific contrast equals zero:

$$H_0 : L = 0 \quad \text{(no difference in the contrast)}$$
$$H_a : L \neq 0 \quad \text{(significant contrast)}$$

We use the test statistic:

$$T = \frac{\hat{L}}{s(\hat{L})} \sim t_{N-a}$$

We reject H_0 at the α level if:

$$|T| > t_{1-\alpha/2;N-a}$$

5.1.1.5.4 Linear Combination of Treatment Means

A **linear combination** of treatment means extends the idea of a contrast:

$$L = \sum_{i=1}^{a} c_i \mu_i$$

Unlike contrasts, there are **no restrictions** on the coefficients c_i (i.e., they do not need to sum to zero).

Since tests on a single treatment mean, pairwise differences, and contrasts are all special cases of this general form, we can express the hypothesis test as:

$$H_0 : \sum_{i=1}^{a} c_i \mu_i = c$$

$$H_a : \sum_{i=1}^{a} c_i \mu_i \neq c$$

The test statistic follows a t-distribution:

$$T = \frac{\hat{L} - c}{s(\hat{L})} \sim t_{N-a}$$

Since squaring a t-distributed variable results in an F-distributed variable,

$$F = T^2 \sim F_{1,N-a}$$

This means that all such tests can be viewed as single-degree-of-freedom F-tests, since the numerator degrees of freedom is always 1.

Multiple Contrasts
When testing $k \geq 2$ contrasts simultaneously, the test statistics $T_1, T_2, ..., T_k$ follow a multivariate t-distribution, since they are dependent (as they are based on the same data).

Limitations of Multiple Comparisons

1. **Inflation of Type I Error**:
   The confidence coefficient $(1 - \alpha)$ applies to a single estimate, not a series of estimates. Similarly, the Type I error rate α applies to an individual test, not a collection of tests.
   Example: If three t-tests are performed at $\alpha = 0.05$, and if they were independent (which they are not), then:

   $$(1 - 0.05)^3 = 0.857$$

   meaning the overall Type I error rate would be approximately 0.143, not 0.05.
2. **Data Snooping Concern**:
   The significance level α is valid only if the test was planned before examining the data.

   - Often, an experiment suggests relationships to investigate.
   - Exploring effects based on observed data is known as data snooping.

To address these issues, we use **Multiple Comparison Procedures**, such as:

- Tukey—for all pairwise comparisons of treatment means.
- Scheffé—for all possible contrasts.
- Bonferroni—for a fixed number of planned comparisons.

5.1.1.5.4.1 Tukey

Used for all pairwise comparisons of treatment means:

$$D = \mu_i - \mu_{i'}$$

Hypothesis test:

$$H_0 : \mu_i - \mu_{i'} = 0$$
$$H_a : \mu_i - \mu_{i'} \neq 0$$

Properties:

- When sample sizes are equal ($n_1 = n_2 = ... = n_a$), the family confidence coefficient is exactly $(1 - \alpha)$.
- When sample sizes are unequal, the method is conservative (i.e., the actual significance level is less than α).

The Tukey test is based on the studentized range:

$$w = \max(Y_i) - \min(Y_i)$$

If $Y_1, ..., Y_r$ are observations from a normal distribution with mean μ and variance σ^2, then the statistic:

$$q(r, v) = \frac{w}{s}$$

follows the studentized range distribution, which requires a special table.

Notes:
- When testing only a subset of pairwise comparisons, the confidence coefficient exceeds $(1 - \alpha)$, making the test more conservative.
- Tukey's method can be used for data snooping, as long as the investigated effects are pairwise comparisons.

5.1.1.5.4.2 Scheffé

Scheffé's method is used for testing **all possible contrasts**:

$$L = \sum_{i=1}^{a} c_i \mu_i, \quad \text{where} \quad \sum_{i=1}^{a} c_i = 0$$

Hypothesis test:

$$H_0 : L = 0$$
$$H_a : L \neq 0$$

Properties:

- Valid for any set of contrasts, making it the most general multiple comparison procedure.
- The family confidence level is exactly $(1 - \alpha)$, regardless of sample sizes.

Simultaneous Confidence Intervals:

$$\hat{L} \pm Ss(\hat{L})$$

where:

- $\hat{L} = \sum c_i \bar{Y}_{i.}$
- $s^2(\hat{L}) = MSE \sum \frac{c_i^2}{n_i}$
- $S^2 = (a - 1) f_{1-\alpha;a-1,N-a}$

Test Statistic:

$$F = \frac{\hat{L}^2}{(a-1)s^2(\hat{L})}$$

We reject H_0 if:

$$F > f_{1-\alpha;a-1,N-a}$$

Notes:

- Finite Family Correction: Since we never test all possible contrasts in practice, the actual confidence coefficient is greater than $(1 - \alpha)$. Thus, some researchers use a higher α (e.g., a 90% confidence level instead of 95%).
- Scheffé is useful for data snooping, since it applies to any contrast.
- If only pairwise comparisons are needed, Tukey's method gives narrower confidence intervals than Scheffé.

5.1.1.5.4.3 Bonferroni

The Bonferroni correction is applicable regardless of whether sample sizes are equal or unequal. It is particularly useful when a small number of planned comparisons are of interest.

A $(1 - \alpha)100\%$ simultaneous confidence interval for a set of g comparisons is:

$$\hat{L} \pm Bs(\hat{L})$$

where:

$$B = t_{1-\alpha/(2g),N-a}$$

and g is the number of comparisons in the family.

To test:

$$H_0 : L = 0$$
$$H_a : L \neq 0$$

we use the test statistic:

$$T = \frac{\hat{L}}{s(\hat{L})}$$

Reject H_0 if:

$$|T| > t_{1-\alpha/(2g),N-a}$$

Notes:

- If all pairwise comparisons are needed, Tukey's method is superior, as it provides narrower confidence intervals.
- Bonferroni is better than Scheffé when the number of contrasts is similar to or smaller than the number of treatment levels.
- Practical recommendation: Compute Tukey, Scheffé, and Bonferroni and use the method with the smallest confidence intervals.
- Bonferroni cannot be used for data snooping, as it assumes the comparisons were planned before examining the data.

5.1.1.5.4.4 Fisher's Least Significant Difference

The Fisher LSD method does not control the family-wise error rate, meaning it does not correct for multiple comparisons. However, it can be useful for exploratory analysis when a preliminary ANOVA is significant.

The hypothesis test for comparing two treatment means:

$$H_0 : \mu_i = \mu_j$$

uses the t-statistic:

$$t = \frac{\bar{Y}_i - \bar{Y}_j}{\sqrt{MSE\left(\frac{1}{n_i} + \frac{1}{n_j}\right)}}$$

where:

- $\bar{Y}_i$ and $\bar{Y}_j$ are the sample means for treatments i and j.
- MSE is the mean squared error from ANOVA.
- n_i, n_j are the sample sizes for groups i and j.

Notes:

- The LSD method does not adjust for multiple comparisons, which increases the Type I error rate.
- It is only valid if the overall ANOVA is significant (i.e., the global null hypothesis of no treatment effect is rejected).
- Tukey and Bonferroni methods are preferred when many comparisons are made.

5.1.1.5.4.5 Newman-Keuls

The Newman-Keuls procedure is a stepwise multiple comparison test similar to Tukey's method but less rigorous.

Key Issues:

- Unlike Tukey, Newman-Keuls does not control the family-wise error rate.
- It has less power than ANOVA.
- It is rarely recommended in modern statistical practice.
- **Do not recommend using the Newman-Keuls test**.

5.1.1.5.4.6 Summary of Multiple Comparison Procedures

5.1.1.5.4.7 Dunnett's Test

In some experiments, instead of comparing all treatment groups against each other, we are specifically interested in comparing each treatment to a control. This is common in clinical trials or A/B testing, where one group serves as a baseline.

Dunnett's test is designed for experiments with a groups, where:

- One group is the control (e.g., placebo or standard treatment).
- The remaining $a - 1$ groups are treatment groups.

Thus, we perform $a - 1$ pairwise comparisons:

$$D_i = \mu_i - \mu_c, \quad i = 1, \ldots, a - 1$$

where μ_c is the mean of the control group.

Dunnett's Test Versus Other Methods

- Unlike Tukey's method (which compares all pairs), Dunnett's method only compares treatments to the control.
- Dunnett's test controls the family-wise error rate, making it more powerful than Bonferroni for this scenario.
- If the goal is to compare treatments against each other as well, Tukey's method is preferable.

Method	Type of comparisons	Controls family-wise error rate?	Best used for	Strengths	Weaknesses
Tukey	All pairwise comparisons	Yes	Comparing all treatment means	Exact confidence level when sample sizes are equal; more powerful than Scheffé for pairwise tests	Conservative if sample sizes are unequal
Scheffé	All possible contrasts	Yes	Exploratory analysis, especially when interested in any contrast	Valid for any contrast; can be used for data snooping	Confidence intervals wider than Tukey for pairwise comparisons
Bonferroni	Fixed number of planned comparisons	Yes	A small number of prespecified tests	Simple and flexible; better than Scheffé for few comparisons	Less powerful than Tukey for many pairwise tests; cannot be used for data snooping
Fisher's LSD	Pairwise comparisons	No	Exploratory comparisons after significant ANOVA	Most powerful for pairwise comparisons when ANOVA is significant	Inflates Type I error rate; not valid without a significant ANOVA
Newman-Keuls	Pairwise comparisons	No	-	-	Less power than ANOVA; generally not recommended

5.1.2 Single-Factor Random Effects ANOVA

Also known as an **ANOVA Type II model**, the single-factor random effects model assumes that treatments are randomly selected from a larger population. Thus, inference extends beyond the observed treatments to the entire population of treatments.

5.1.2.1 Random Cell Means Model

The model is given by:

$$Y_{ij} = \mu_i + \epsilon_{ij}$$

where:

- $\mu_i \sim N(\mu, \sigma_\mu^2)$, independent across treatments.
- $\epsilon_{ij} \sim N(0, \sigma^2)$, independent across observations.
- μ_i and ϵ_{ij} are mutually independent for $i = 1, \ldots, a$ and $j = 1, \ldots, n$.

When all treatment sample sizes are equal:

$$
\begin{aligned}
E(Y_{ij}) &= E(\mu_i) = \mu \\
var(Y_{ij}) &= var(\mu_i) + var(\epsilon_{ij}) = \sigma_\mu^2 + \sigma^2
\end{aligned}
$$

5.1.2.1.1 Covariance Structure

Since Y_{ij} are not independent, we calculate their covariances:

1. **Same treatment group** (i fixed, $j \neq j'$):

$$
\begin{aligned}
cov(Y_{ij}, Y_{ij'}) &= E(Y_{ij}Y_{ij'}) - E(Y_{ij})E(Y_{ij'}) \\
&= E(\mu_i^2 + \mu_i\epsilon_{ij'} + \mu_i\epsilon_{ij} + \epsilon_{ij}\epsilon_{ij'}) - \mu^2 \\
&= \sigma_\mu^2 + \mu^2 - \mu^2 \\
&= \sigma_\mu^2
\end{aligned}
$$

2. **Different treatment groups** ($i \neq i'$):

$$
\begin{aligned}
cov(Y_{ij}, Y_{i'j'}) &= E(\mu_i\mu_{i'} + \mu_i\epsilon_{i'j'} + \mu_{i'}\epsilon_{ij} + \epsilon_{ij}\epsilon_{i'j'}) - \mu^2 \\
&= \mu^2 - \mu^2 = 0
\end{aligned}
$$

Thus:

- All observations have the same variance: $var(Y_{ij}) = \sigma_\mu^2 + \sigma^2$.
- Observations from the same treatment have covariance: σ_μ^2.
- Observations from different treatments are uncorrelated.

The **intraclass correlation** between two responses from the same treatment:

$$
\rho(Y_{ij}, Y_{ij'}) = \frac{\sigma_\mu^2}{\sigma_\mu^2 + \sigma^2}, \quad j \neq j'
$$

5.1.2.1.2 Inference for Random Effects Model

The **Intraclass Correlation Coefficient**:

$$\frac{\sigma_\mu^2}{\sigma^2 + \sigma_\mu^2}$$

measures the proportion of total variability in Y_{ij} that is accounted for by treatment differences.

To test whether treatments contribute significantly to variance:

$$H_0 : \sigma_\mu^2 = 0 \quad \text{(No treatment effect, all } \mu_i = \mu\text{)}$$
$$H_a : \sigma_\mu^2 \neq 0$$

Under H_0, an ANOVA F-test is used:

$$F = \frac{MSTR}{MSE}$$

where:

- $MSTR$ (Mean Square for Treatments) captures variation **between treatments**.
- MSE (Mean Square Error) captures variation **within treatments**.

If H_0 is true, then:

$$F \sim F_{(a-1,a(n-1))}$$

Reject H_0 if:

$$F > f_{(1-\alpha;a-1,a(n-1))}$$

5.1.2.1.3 Comparison: Fixed Effects Versus Random Effects Models

Although ANOVA calculations are the same for fixed and random effects models, the interpretation of results differs (Table 5.4).

- If $\sigma_\mu^2 = 0$, then $E(MSTR) = E(MSE)$, implying no treatment effect.
- Otherwise, $E(MSTR) > E(MSE)$, suggesting significant treatment variation.

Table 5.4 Expected mean squares under fixed and random effects models

Random effects model	Fixed effects model
$E(MSE) = \sigma^2$	$E(MSE) = \sigma^2$
$E(MSTR) = \sigma^2 + n\sigma_\mu^2$	$E(MSTR) = \sigma^2 + \frac{\sum_i n_i(\mu_i-\mu)^2}{a-1}$

When sample sizes are not equal, the F-test remains valid, but the degrees of freedom change to:

$$F \sim F_{(a-1,N-a)}$$

5.1.2.1.4 Estimation of μ

An unbiased estimator of $E(Y_{ij}) = \mu$ is the grand mean:

$$\hat{\mu} = \bar{Y}_{..} = \frac{1}{an} \sum_{i=1}^{a} \sum_{j=1}^{n} Y_{ij}$$

The variance of this estimator is:

$$\begin{aligned} var(\bar{Y}_{..}) &= var\left(\frac{1}{a}\sum_{i=1}^{a} \bar{Y}_{i.}\right) \\ &= \frac{1}{a^2}\sum_{i=1}^{a} var(\bar{Y}_{i.}) \\ &= \frac{1}{a^2}\sum_{i=1}^{a}\left(\sigma_{\mu}^2 + \frac{\sigma^2}{n}\right) \\ &= \frac{n\sigma_{\mu}^2 + \sigma^2}{an} \end{aligned}$$

An unbiased estimator of this variance is:

$$s^2(\bar{Y}_{..}) = \frac{MSTR}{an}$$

Since:

$$\frac{\bar{Y}_{..} - \mu}{s(\bar{Y}_{..})} \sim t_{a-1}$$

A $(1-\alpha)100\%$ confidence interval for μ is:

$$\bar{Y}_{..} \pm t_{1-\alpha/2;a-1} s(\bar{Y}_{..})$$

5.1.2.1.5 Estimation of Intraclass Correlation Coefficient $\frac{\sigma_\mu^2}{\sigma_\mu^2+\sigma^2}$

In both random and fixed effects models, $MSTR$ and MSE are independent.

When sample sizes are equal ($n_i = n$ for all i), the test statistic:

$$\frac{\frac{MSTR}{n\sigma_\mu^2+\sigma^2}}{\frac{MSE}{\sigma^2}} \sim F_{a-1,a(n-1)}$$

A $(1-\alpha)100\%$ confidence interval for $\frac{\sigma_\mu^2}{\sigma_\mu^2+\sigma^2}$ follows from:

$$P\left(f_{\alpha/2;a-1,a(n-1)} \le \frac{\frac{MSTR}{n\sigma_\mu^2+\sigma^2}}{\frac{MSE}{\sigma^2}} \le f_{1-\alpha/2;a-1,a(n-1)}\right) = 1-\alpha$$

Defining:

$$L = \frac{1}{n}\left(\frac{MSTR}{MSE} \times \frac{1}{f_{1-\alpha/2;a-1,a(n-1)}} - 1\right)$$
$$U = \frac{1}{n}\left(\frac{MSTR}{MSE} \times \frac{1}{f_{\alpha/2;a-1,a(n-1)}} - 1\right)$$

The lower and upper confidence limits for $\frac{\sigma_\mu^2}{\sigma_\mu^2+\sigma^2}$ are:

$$L^* = \frac{L}{1+L}$$
$$U^* = \frac{U}{1+U}$$

If L^* is negative, we customarily set it to 0.

5.1.2.1.6 Estimation of σ^2

Since:

$$\frac{a(n-1)MSE}{\sigma^2} \sim \chi^2_{a(n-1)}$$

A $(1-\alpha)100\%$ confidence interval for σ^2 is:

$$\frac{a(n-1)MSE}{\chi^2_{1-\alpha/2;a(n-1)}} \le \sigma^2 \le \frac{a(n-1)MSE}{\chi^2_{\alpha/2;a(n-1)}}$$

If sample sizes are unequal, the same formula applies, but the degrees of freedom change to:

$$df = N - a$$

5.1.2.1.7 Estimation of σ_μ^2

From the expectations:

$$E(MSE) = \sigma^2, \quad E(MSTR) = \sigma^2 + n\sigma_\mu^2$$

we solve for σ_μ^2:

$$\sigma_\mu^2 = \frac{E(MSTR) - E(MSE)}{n}$$

An unbiased estimator of σ_μ^2 is:

$$s_\mu^2 = \frac{MSTR - MSE}{n}$$

If $s_\mu^2 < 0$, we set $s_\mu^2 = 0$ (since variances cannot be negative).
If sample sizes are unequal, we replace n with an effective sample size n':

$$s_\mu^2 = \frac{MSTR - MSE}{n'}$$

where:

$$n' = \frac{1}{a-1}\left(\sum_i n_i - \frac{\sum_i n_i^2}{\sum_i n_i}\right)$$

There are no exact confidence intervals for σ_μ^2, but we can approximate them using the Satterthwaite procedure.

5.1.2.1.7.1 Satterthwaite Approximation

A linear combination of expected mean squares:

$$\sigma_\mu^2 = \frac{1}{n}E(MSTR) + \left(-\frac{1}{n}\right)E(MSE)$$

For a general linear combination:

$$S = d_1 E(MS_1) + \cdots + d_h E(MS_h)$$

where d_i are coefficients, an unbiased estimator of S is:

$$\hat{S} = d_1 MS_1 + \cdots + d_h MS_h$$

Let df_i be the degrees of freedom associated with each mean square MS_i. The Satterthwaite approximation states:

$$\frac{(df)\hat{S}}{S} \sim \chi^2_{df}$$

where the degrees of freedom are approximated as:

$$df = \frac{(d_1 MS_1 + \cdots + d_h MS_h)^2}{\sum_{i=1}^{h} \frac{(d_i MS_i)^2}{df_i}}$$

Applying the Satterthwaite method to the single-factor random effects model:

$$\frac{(df)s^2_\mu}{\chi^2_{1-\alpha/2;df}} \leq \sigma^2_\mu \leq \frac{(df)s^2_\mu}{\chi^2_{\alpha/2;df}}$$

where the approximate degrees of freedom are:

$$df = \frac{(s^2_\mu)^2}{\frac{(MSTR)^2}{a-1} + \frac{(MSE)^2}{a(n-1)}}$$

5.1.2.2 Random Treatment Effects Model

In a random effects model, treatment levels are considered random samples from a larger population of possible treatments. The model accounts for variability across all potential treatments, not just those observed in the study.

We define the **random treatment effect** as:

$$\tau_i = \mu_i - E(\mu_i) = \mu_i - \mu$$

where τ_i represents the deviation of treatment mean μ_i from the overall mean μ.

Thus, we rewrite treatment means as:

$$\mu_i = \mu + \tau_i$$

Table 5.5 Common regression issues and diagnostic tools

Issue	Diagnostic tools
Nonconstant error variance (heteroscedasticity)	Residual plots, Levene's test, Hartley's test
Nonindependence of errors	Residual plots, Durbin-Watson test (for autocorrelation)
Outliers	Boxplots, residual plots, regression influence measures (e.g., Cook's distance)
Non-normality of errors	Histogram, Q-Q plot, Shapiro-Wilk test, Anderson-Darling test
Omitted variable bias	Residual plots, checking for unaccounted sources of variation

Substituting this into the response model:

$$Y_{ij} = \mu + \tau_i + \epsilon_{ij}$$

where:

- μ = common mean across all observations.
- $\tau_i \sim N(0, \sigma_\tau^2)$, random treatment effects, assumed independent.
- $\epsilon_{ij} \sim N(0, \sigma^2)$, random error terms, also independent.
- τ_i and ϵ_{ij} are mutually independent for $i = 1, \ldots, a$ and $j = 1, \ldots, n$.
- We consider only balanced single-factor ANOVA (equal sample sizes across treatments).

5.1.2.3 Diagnostic Measures for Model Assumptions

Checking assumptions is crucial for valid inference. Common issues include (Table 5.5):

5.1.2.4 Remedial Measures

If diagnostic checks indicate violations of assumptions, possible solutions include:

- Weighted Least Squares—Adjusts for heteroscedasticity.
- Variable Transformation—Log or Box-Cox transformations may improve normality or stabilize variance.

- Nonparametric Procedures—Kruskal-Wallis test or bootstrapping when normality assumptions fail.

5.1.2.5 Key Notes on Robustness

- Fixed effects ANOVA is relatively robust to:
  - Non-normality, particularly when sample sizes are moderate to large.
  - Unequal variances when sample sizes are roughly equal.
  - F-test and multiple comparisons remain valid under mild violations.
- Random effects ANOVA is sensitive to:
  - Lack of independence, which severely affects both fixed and random effects models.
  - Unequal variances, particularly when estimating variance components.

5.1.3 Two-Factor Fixed Effects ANOVA

A multifactor experiment offers several advantages:

- Higher efficiency—More precise estimates with fewer observations.
- Increased information—Allows for testing interactions between factors.
- Greater validity—Reduces confounding by controlling additional sources of variation.

Balanced Two-Factor ANOVA: Assumptions

- Equal sample sizes for all treatment combinations.
- All treatment means are of equal importance (no weighting).
- Factors are categorical and chosen purposefully.

We assume:

- Factor A has a levels and Factor B has b levels.
- All $a \times b$ factor level combinations are included.
- Each treatment combination has n replications.
- The total number of observations:

$$N = abn$$

5.1.3.1 Cell Means Model

The response is modeled as:

$$Y_{ijk} = \mu_{ij} + \epsilon_{ijk}$$

where:

- μ_{ij} are fixed parameters (cell means).
- $i = 1, \ldots, a$ represents levels of Factor A.
- $j = 1, \ldots, b$ represents levels of Factor B.
- $\epsilon_{ijk} \sim$ independent $N(0, \sigma^2)$ for all i, j, k.

  Expected values and variance:

$$E(Y_{ijk}) = \mu_{ij}$$
$$var(Y_{ijk}) = var(\epsilon_{ijk}) = \sigma^2$$

  Thus:

$$Y_{ijk} \sim \text{independent } N(\mu_{ij}, \sigma^2)$$

  This can be expressed in matrix notation:

$$\mathbf{Y} = \mathbf{X}\beta + \epsilon$$

where:

$$E(\mathbf{Y}) = \mathbf{X}\beta$$
$$var(\mathbf{Y}) = \sigma^2\mathbf{I}$$

5.1.3.1.1 Interaction Effects

Interaction measures whether the effect of one factor depends on the level of the other factor. It is defined as:

$$(\alpha\beta)_{ij} = \mu_{ij} - (\mu_{..} + \alpha_i + \beta_j)$$

where:

- **Grand mean**:

$$\mu_{..} = \frac{1}{ab}\sum_i \sum_j \mu_{ij}$$

- **Main effect for Factor A** (average effect of level i):

$$\alpha_i = \mu_{i.} - \mu_{..}$$

- **Main effect for Factor B** (average effect of level j):

$$\beta_j = \mu_{.j} - \mu_{..}$$

- **Interaction effect**:

$$(\alpha\beta)_{ij} = \mu_{ij} - \mu_{i.} - \mu_{.j} + \mu_{..}$$

To determine whether interactions exist:

1. Check if all μ_{ij} can be written as sums $\mu_{..} + \alpha_i + \beta_j$ (i.e., check if interaction terms are zero).
2. Compare mean differences across levels of Factor B at each level of Factor A.
3. Compare mean differences across levels of Factor A at each level of Factor B.
4. Graphical method:
   - Plot treatment means for each level of Factor B.
   - If lines are not parallel, an interaction exists.

The interaction terms satisfy:
For each level of **Factor B**:

$$\sum_i (\alpha\beta)_{ij} = \sum_i \left(\mu_{ij} - \mu_{..} - \alpha_i - \beta_j\right)$$

Expanding:

$$\begin{aligned}\sum_i (\alpha\beta)_{ij} &= \sum_i \mu_{ij} - a\mu_{..} - \sum_i \alpha_i - a\beta_j \\ &= a\mu_{.j} - a\mu_{..} - \sum_i (\mu_{i.} - \mu_{..}) - a(\mu_{.j} - \mu_{..}) \\ &= a\mu_{.j} - a\mu_{..} - a\mu_{..} + a\mu_{..} - a(\mu_{.j} - \mu_{..}) \\ &= 0\end{aligned}$$

Similarly:

$$\sum_j (\alpha\beta)_{ij} = 0, \quad i = 1, \ldots, a$$

and:

$$\sum_i \sum_j (\alpha\beta)_{ij} = 0, \quad \sum_i \alpha_i = 0, \quad \sum_j \beta_j = 0$$

5.1.3.2 Factor Effects Model

In the Factor Effects Model, we express the response as:

$$\mu_{ij} = \mu_{..} + \alpha_i + \beta_j + (\alpha\beta)_{ij}$$
$$Y_{ijk} = \mu_{..} + \alpha_i + \beta_j + (\alpha\beta)_{ij} + \epsilon_{ijk}$$

where:

- $\mu_{..}$ is the grand mean.
- α_i are main effects for Factor A, subject to:

$$\sum_i \alpha_i = 0$$

- β_j are main effects for Factor B, subject to:

$$\sum_j \beta_j = 0$$

- $(\alpha\beta)_{ij}$ are interaction effects, subject to:

$$\sum_i (\alpha\beta)_{ij} = 0, \quad j = 1, \ldots, b$$

$$\sum_j (\alpha\beta)_{ij} = 0, \quad i = 1, \ldots, a$$

- $\epsilon_{ijk} \sim$ independent $N(0, \sigma^2)$ for $k = 1, \ldots, n$.

  Thus, we have:

$$E(Y_{ijk}) = \mu_{..} + \alpha_i + \beta_j + (\alpha\beta)_{ij}$$
$$var(Y_{ijk}) = \sigma^2$$
$$Y_{ijk} \sim N(\mu_{..} + \alpha_i + \beta_j + (\alpha\beta)_{ij}, \sigma^2)$$

Table 5.6 Parameter count in Two-Way ANOVA with interaction

Parameter	Count
$\mu_{..}$	1
α_i (Main effects for A)	$a-1$ (due to constraint $\sum_i \alpha_i = 0$)
β_j (Main effects for B)	$b-1$ (due to constraint $\sum_j \beta_j = 0$)
$(\alpha\beta)_{ij}$ (Interaction effects)	$(a-1)(b-1)$ (due to two constraints)

5.1.3.3 Parameter Counting and Restrictions

The Cell Means Model has ab parameters corresponding to each combination of factor levels.

In the Factor Effects Model, the imposed constraints reduce the number of estimable parameters (Table 5.6).

Thus, the total number of parameters:

$$1 + (a-1) + (b-1) + (a-1)(b-1) = ab$$

which matches the number of parameters in the Cell Means Model.

To uniquely estimate parameters, we apply constraints:

$$\begin{aligned}
\alpha_a &= -(\alpha_1 + \alpha_2 + \cdots + \alpha_{a-1}) \\
\beta_b &= -(\beta_1 + \beta_2 + \cdots + \beta_{b-1}) \\
(\alpha\beta)_{ib} &= -(\alpha\beta)_{i1} - (\alpha\beta)_{i2} - \cdots - (\alpha\beta)_{i,b-1}, \quad i = 1, \ldots, a \\
(\alpha\beta)_{aj} &= -(\alpha\beta)_{1j} - (\alpha\beta)_{2j} - \cdots - (\alpha\beta)_{a-1,j}, \quad j = 1, \ldots, b
\end{aligned}$$

The model can be fitted using least squares or maximum likelihood estimation.

5.1.3.3.1 Cell Means Model Estimation

Minimizing:

$$Q = \sum_i \sum_j \sum_k (Y_{ijk} - \mu_{ij})^2$$

yields estimators:

$$\begin{aligned}
\hat{\mu}_{ij} &= \bar{Y}_{ij} \\
\hat{Y}_{ijk} &= \bar{Y}_{ij} \\
e_{ijk} &= Y_{ijk} - \hat{Y}_{ijk} = Y_{ijk} - \bar{Y}_{ij}
\end{aligned}$$

where $e_{ijk} \sim$ independent $N(0, \sigma^2)$.

5.1.3.3.2 Factor Effects Model Estimation

Minimizing:

$$Q = \sum_i \sum_j \sum_k (Y_{ijk} - \mu_{..} - \alpha_i - \beta_j - (\alpha\beta)_{ij})^2$$

subject to the constraints:

$$\sum_i \alpha_i = 0$$
$$\sum_j \beta_j = 0$$
$$\sum_i (\alpha\beta)_{ij} = 0, \quad j = 1, \ldots, b$$
$$\sum_j (\alpha\beta)_{ij} = 0, \quad i = 1, \ldots, a$$

yields estimators:

$$\hat{\mu}_{..} = \bar{Y}_{...}$$
$$\hat{\alpha}_i = \bar{Y}_{i..} - \bar{Y}_{...}$$
$$\hat{\beta}_j = \bar{Y}_{.j.} - \bar{Y}_{...}$$
$$(\hat{\alpha\beta})_{ij} = \bar{Y}_{ij.} - \bar{Y}_{i..} - \bar{Y}_{.j.} + \bar{Y}_{...}$$

The fitted values are:

$$\hat{Y}_{ijk} = \bar{Y}_{...} + (\bar{Y}_{i..} - \bar{Y}_{...}) + (\bar{Y}_{.j.} - \bar{Y}_{...}) + (\bar{Y}_{ij.} - \bar{Y}_{i..} - \bar{Y}_{.j.} + \bar{Y}_{...})$$

which simplifies to:

$$\hat{Y}_{ijk} = \bar{Y}_{ij.}$$

The residuals are:

$$e_{ijk} = Y_{ijk} - \bar{Y}_{ij.}$$

and follow:

$$e_{ijk} \sim \text{independent } N(0, \sigma^2)$$

The variances of the estimated effects are:

$$s^2_{\hat{\mu}_{..}} = \frac{MSE}{nab}$$

$$s^2_{\hat{\alpha}_i} = MSE\left(\frac{1}{nb} - \frac{1}{nab}\right)$$

$$s^2_{\hat{\beta}_j} = MSE\left(\frac{1}{na} - \frac{1}{nab}\right)$$

$$s^2_{(\hat{\alpha\beta})_{ij}} = MSE\left(\frac{1}{n} - \frac{1}{na} - \frac{1}{nb} + \frac{1}{nab}\right)$$

5.1.3.3.3 Partitioning the Total Sum of Squares

The total deviation of an observation from the overall mean can be decomposed as:

$$Y_{ijk} - \bar{Y}_{...} = (\bar{Y}_{ij.} - \bar{Y}_{...}) + (Y_{ijk} - \bar{Y}_{ij.})$$

where:

- $Y_{ijk} - \bar{Y}_{...}$: Total deviation of an observation.
- $\bar{Y}_{ij.} - \bar{Y}_{...}$: Deviation of treatment mean from the overall mean.
- $Y_{ijk} - \bar{Y}_{ij.}$: Residual deviation of an observation from the treatment mean.

Summing over all observations:

$$\sum_i \sum_j \sum_k (Y_{ijk} - \bar{Y}_{...})^2 = n \sum_i \sum_j (\bar{Y}_{ij.} - \bar{Y}_{...})^2 + \sum_i \sum_j \sum_k (Y_{ijk} - \bar{Y}_{ij.})^2$$

Thus:

$$SSTO = SSTR + SSE$$

where:

- $SSTO$ = Total Sum of Squares (Total variation).
- $SSTR$ = Treatment Sum of Squares (Variation due to factor effects).
- SSE = Error Sum of Squares (Residual variation).

Since the cross-product terms are 0, the model naturally partitions the variance. From the factor effects model:

$$\bar{Y}_{ij.} - \bar{Y}_{...} = (\bar{Y}_{i..} - \bar{Y}_{...}) + (\bar{Y}_{.j.} - \bar{Y}_{...}) + (\bar{Y}_{ij.} - \bar{Y}_{i..} - \bar{Y}_{.j.} + \bar{Y}_{...})$$

Squaring and summing:

$$n\sum_i\sum_j(\bar{Y}_{ij.} - \bar{Y}_{...})^2 = nb\sum_i(\bar{Y}_{i..} - \bar{Y}_{...})^2 + na\sum_j(\bar{Y}_{.j.} - \bar{Y}_{...})^2$$
$$+n\sum_i\sum_j(\bar{Y}_{ij.} - \bar{Y}_{i..} - \bar{Y}_{.j.} + \bar{Y}_{...})^2$$

Thus, treatment sum of squares can be further partitioned as:

$$SSTR = SSA + SSB + SSAB$$

where:

- SSA: Sum of Squares for Factor A.
- SSB: Sum of Squares for Factor B.
- $SSAB$: Sum of Squares for Interaction.

The interaction term can also be expressed as:

$$SSAB = SSTO - SSE - SSA - SSB$$

or equivalently:

$$SSAB = SSTR - SSA - SSB$$

where:

- SSA measures the variability of the estimated factor **A** level means ($\bar{Y}_{i..}$). The more variable these means, the larger SSA.
- SSB measures the variability of the estimated factor **B** level means ($\bar{Y}_{.j.}$).
- $SSAB$ measures the variability in **interaction effects**.

For Two-Factor ANOVA, the degrees of freedom partitioning follows (Table 5.7):
Since:

$$SSTR = SSA + SSB + SSAB$$

Table 5.7 Sum of squares and associated degrees of freedom

Sum of squares	Degrees of freedom (df)
$SSTO$ (Total)	$N - 1 = abn - 1$
$SSTR$ (Treatments)	$ab - 1$
SSE (Error)	$N - ab = ab(n - 1)$
SSA (Factor A)	$a - 1$
SSB (Factor B)	$b - 1$
$SSAB$ (Interaction)	$(a - 1)(b - 1)$

the treatment degrees of freedom also partition as:

$$ab - 1 = (a - 1) + (b - 1) + (a - 1)(b - 1)$$

- $df_{SSA} = a - 1$
  (One degree of freedom lost due to the constraint $\sum(\bar{Y}_{i..} - \bar{Y}_{...}) = 0$).
- $df_{SSB} = b - 1$
  (One degree of freedom lost due to the constraint $\sum(\bar{Y}_{.j.} - \bar{Y}_{...}) = 0$).
- $df_{SSAB} = (a - 1)(b - 1)$
  (Due to interaction constraints).

The Mean Squares are obtained by dividing Sum of Squares by the corresponding degrees of freedom:

$$MSA = \frac{SSA}{a - 1}$$

$$MSB = \frac{SSB}{b - 1}$$

$$MSAB = \frac{SSAB}{(a - 1)(b - 1)}$$

The expectations of the mean squares are:

$$E(MSE) = \sigma^2$$

$$E(MSA) = \sigma^2 + nb\frac{\sum \alpha_i^2}{a - 1} = \sigma^2 + nb\frac{\sum(\mu_{i..} - \mu_{..})^2}{a - 1}$$

$$E(MSB) = \sigma^2 + na\frac{\sum \beta_j^2}{b - 1} = \sigma^2 + na\frac{\sum(\mu_{.j.} - \mu_{..})^2}{b - 1}$$

$$E(MSAB) = \sigma^2 + n\frac{\sum\sum(\alpha\beta)_{ij}^2}{(a - 1)(b - 1)} = \sigma^2 + n\frac{\sum(\mu_{ij} - \mu_{i..} - \mu_{.j.} + \mu_{..})^2}{(a - 1)(b - 1)}$$

If Factor A has no effect ($\mu_{i..} = \mu_{..}$), then MSA and MSE have the same expectation.

Similarly, if Factor B has no effect, then $MSB = MSE$.

Thus, MSA > MSE and MSB > MSE suggest the presence of factor effects.

5.1.3.4 Testing for Interaction

Hypotheses:

$$H_0 : \mu_{ij} - \mu_{i..} - \mu_{.j.} + \mu_{..} = 0 \quad \text{(No interaction)}$$

$$H_a : \mu_{ij} - \mu_{i..} - \mu_{.j.} + \mu_{..} \neq 0 \quad \text{(Interaction present)}$$

or equivalently:

$$H_0 : \text{All } (\alpha\beta)_{ij} = 0$$
$$H_a : \text{Not all } (\alpha\beta)_{ij} = 0$$

The F-statistic is:

$$F = \frac{MSAB}{MSE}$$

Under H_0, $F \sim F_{(a-1)(b-1),ab(n-1)}$. Reject H_0 if:

$$F > F_{1-\alpha;(a-1)(b-1),ab(n-1)}$$

5.1.3.5 Two-Way ANOVA Summary Table

The Two-Way ANOVA table partitions the total variation into its components (Table 5.8):

Interpreting Two-Way ANOVA Results

When conducting a Two-Way ANOVA, always check interaction effects first:

1. If the interaction $(A \times B)$ is significant:
   - The effect of one factor depends on the level of the other factor.
   - Main effects are not interpretable alone because their impact varies across levels of the second factor.
2. If the interaction is NOT significant:
   - The factors have independent (additive) effects.
   - Main effects can be tested individually.

Table 5.8 Source of variation, sums of squares, and F-statistics

Source of variation	Sum of squares (SS)	Degrees of freedom (df)	Mean square (MS)	F-Statistic
Factor A	SSA	$a-1$	$MSA = \frac{SSA}{a-1}$	$F_A = \frac{MSA}{MSE}$
Factor B	SSB	$b-1$	$MSB = \frac{SSB}{b-1}$	$F_B = \frac{MSB}{MSE}$
Interaction (A × B)	$SSAB$	$(a-1)(b-1)$	$MSAB = \frac{SSAB}{(a-1)(b-1)}$	$F_{AB} = \frac{MSAB}{MSE}$
Error	SSE	$ab(n-1)$	$MSE = \frac{SSE}{ab(n-1)}$	–
Total (corrected)	$SSTO$	$abn-1$	–	–

Post Hoc Comparisons

- If interaction is not significant, proceed with main effect comparisons using:
  - Tukey
  - Scheffé
  - Bonferroni
- If interaction is significant, post hoc tests should examine simple effects (comparisons within each level of a factor).

5.1.3.5.1 Contrasts in Two-Way ANOVA

In Two-Way ANOVA, we can define contrasts to test specific hypotheses:

$$L = \sum c_i \mu_i, \quad \text{where } \sum c_i = 0$$

An unbiased estimator of L:

$$\hat{L} = \sum c_i \bar{Y}_{i..}$$

with variance:

$$\sigma^2(\hat{L}) = \frac{\sigma^2}{bn} \sum c_i^2$$

and variance estimate:

$$\frac{MSE}{bn} \sum c_i^2$$

5.1.3.5.1.1 Orthogonal Contrasts in Two-Way ANOVA

For two contrasts:

$$L_1 = \sum c_i \mu_i, \quad \sum c_i = 0$$

$$L_2 = \sum d_i \mu_i, \quad \sum d_i = 0$$

They are orthogonal if:

$$\sum \frac{c_i d_i}{n_i} = 0$$

For balanced designs ($n_i = n$):

$$\sum c_i d_i = 0$$

This ensures that orthogonal contrasts are uncorrelated:

$$\begin{aligned} cov(\hat{L}_1, \hat{L}_2) &= cov\left(\sum_i c_i \bar{Y}_{i..}, \sum_l d_l \bar{Y}_{l..}\right) \\ &= \sum_i \sum_l c_i d_l cov(\bar{Y}_{i..}, \bar{Y}_{l..}) \\ &= \sum_i c_i d_i \frac{\sigma^2}{bn} = 0 \end{aligned}$$

Thus, orthogonal contrasts allow us to partition the sum of squares further.

5.1.3.5.1.2 Orthogonal Polynomial Contrasts

- Used when factor levels are equally spaced (e.g., dose levels: 0, 15, 30, 45, 60).
- Requires equal sample sizes across factor levels.

The Sum of Squares (SS) for a given contrast:

$$SS_L = \frac{\hat{L}^2}{\sum_{i=1}^{a} \frac{c_i^2}{bn_i}}$$

The t-statistic for testing contrasts:

$$T = \frac{\hat{L}}{\sqrt{MSE \sum_{i=1}^{a} \frac{c_i^2}{bn_i}}} \sim t$$

Since:

$$t^2_{(1-\alpha/2;df)} = F_{(1-\alpha;1,df)}$$

we can equivalently test:

$$\frac{SS_L}{MSE} \sim F_{(1-\alpha;1,df_{MSE})}$$

All contrasts have $df = 1$.

5.1.3.6 Unbalanced Two-Way ANOVA

In many practical situations, sample sizes may be unequal across factor combinations, such as in:

- Observational studies (e.g., real-world data with missing values).
- Dropouts in designed studies (e.g., clinical trials with subject attrition).
- Larger sample sizes for inexpensive treatments.
- Sample sizes chosen to match population proportions.

We assume the standard Two-Way ANOVA model:

$$Y_{ijk} = \mu_{..} + \alpha_i + \beta_j + (\alpha\beta)_{ij} + \epsilon_{ijk}$$

where sample sizes vary:

$$n_{i.} = \sum_j n_{ij} \quad \text{(Total for factor level } i\text{)}$$

$$n_{.j} = \sum_i n_{ij} \quad \text{(Total for factor level } j\text{)}$$

$$n_T = \sum_i \sum_j n_{ij} \quad \text{(Total sample size)}$$

However, for unbalanced designs, a major issue arises:

$$SSTO \neq SSA + SSB + SSAB + SSE$$

Unlike the balanced case, the design is non-orthogonal, meaning sum-of-squares partitions do not add up cleanly.

5.1.3.6.1 Indicator Variables for Factor Levels

To handle unbalanced data, we use indicator (dummy) variables as predictors.

For Factor A ($i = 1, \ldots, a-1$):

$$u_i = \begin{cases} +1 & \text{if observation is from level } i \text{ of Factor A} \\ -1 & \text{if observation is from the reference level (level } a\text{)} \\ 0 & \text{otherwise} \end{cases}$$

For Factor B ($j = 1, \ldots, b-1$):

$$v_j = \begin{cases} +1 & \text{if observation is from level } j \text{ of Factor B} \\ -1 & \text{if observation is from the reference level (level } b) \\ 0 & \text{otherwise} \end{cases}$$

Rewriting the ANOVA model using indicator variables:

$$Y = \mu_{..} + \sum_{i=1}^{a-1} \alpha_i u_i + \sum_{j=1}^{b-1} \beta_j v_j + \sum_{i=1}^{a-1} \sum_{j=1}^{b-1} (\alpha\beta)_{ij} u_i v_j + \epsilon$$

Here, the unknown parameters are:

- $\mu_{..}$ (grand mean)
- α_i (main effects for Factor A)
- β_j (main effects for Factor B)
- $(\alpha\beta)_{ij}$ (interaction effects)

5.1.3.6.2 Hypothesis Testing Using Extra Sum of Squares

For unbalanced designs, we use sequential (type I) or adjusted (type III) sum of squares to test hypotheses.

To test for interaction effects, we test:

$$H_0 : \text{All } (\alpha\beta)_{ij} = 0 \quad \text{(No interaction)}$$
$$H_a : \text{Not all } (\alpha\beta)_{ij} = 0 \quad \text{(Interaction present)}$$

To test whether Factor B has an effect:

$$H_0 : \beta_1 = \beta_2 = \cdots = \beta_b = 0$$
$$H_a : \text{At least one } \beta_j \neq 0$$

5.1.3.6.3 Factor Mean Analysis and Contrasts

Factor means and contrasts (e.g., pairwise comparisons) work similarly to the balanced case but require adjustments due to unequal sample sizes.

The variance estimate for a contrast:

$$\sigma^2(\hat{L}) = \frac{\sigma^2}{\sum n_{ij}} \sum c_i^2$$

is modified to:

$$\frac{MSE}{\sum n_{ij}} \sum c_i^2$$

Orthogonal contrasts are harder to define because unequal sample sizes break orthogonality.

5.1.3.6.4 Regression Approach to Unbalanced ANOVA

An alternative is to fit the cell means model as a regression model:

$$Y_{ij} = \mu_{ij} + \epsilon_{ij}$$

which allows us to analyze each treatment mean separately.

However, if there are empty cells (some factor combinations have no observations), the regression approach fails, and only partial analyses can be conducted.

5.1.4 Two-Way Random Effects ANOVA

The Two-Way Random Effects ANOVA assumes that both Factor A and Factor B levels are randomly sampled from larger populations.

The model is:

$$Y_{ijk} = \mu_{..} + \alpha_i + \beta_j + (\alpha\beta)_{ij} + \epsilon_{ijk}$$

where:

- $\mu_{..}$: Overall mean (constant).
- $\alpha_i \sim N(0, \sigma_\alpha^2)$ for $i = 1, \ldots, a$ (random effects for Factor A, independently distributed).
- $\beta_j \sim N(0, \sigma_\beta^2)$ for $j = 1, \ldots, b$ (random effects for Factor B, independently distributed).
- $(\alpha\beta)_{ij} \sim N(0, \sigma_{\alpha\beta}^2)$ for $i = 1, \ldots, a$, $j = 1, \ldots, b$ (random interaction effects, independently distributed).
- $\epsilon_{ijk} \sim N(0, \sigma^2)$ (random error, independently distributed).

Additionally, all random effects $(\alpha_i, \beta_j, (\alpha\beta)_{ij})$ and error terms (ϵ_{ijk}) are mutually independent.

5.1.4.1 Expectation

Taking expectations on both sides:

$$E(Y_{ijk}) = E(\mu_{..} + \alpha_i + \beta_j + (\alpha\beta)_{ij} + \epsilon_{ijk})$$

Since all random effects have mean zero:

$$E(Y_{ijk}) = \mu_{..}$$

Thus, the mean response across all factor levels is $\mu_{...}$.

5.1.4.2 Variance

The total variance of observations is the sum of all variance components:

$$\begin{aligned} var(Y_{ijk}) &= var(\alpha_i) + var(\beta_j) + var((\alpha\beta)_{ij}) + var(\epsilon_{ijk}) \\ &= \sigma_\alpha^2 + \sigma_\beta^2 + \sigma_{\alpha\beta}^2 + \sigma^2 \end{aligned}$$

Thus:

$$Y_{ijk} \sim N(\mu_{..}, \sigma_\alpha^2 + \sigma_\beta^2 + \sigma_{\alpha\beta}^2 + \sigma^2)$$

5.1.4.3 Covariance Structure

In random effects models, observations are correlated if they share the same factor levels.

Case 1: Same Factor A, Different Factor B
If i is the same but $j \neq j'$, then:

$$cov(Y_{ijk}, Y_{ij'k'}) = var(\alpha_i) = \sigma_\alpha^2$$

Case 2: Same Factor B, Different Factor A
If j is the same but $i \neq i'$, then:

$$cov(Y_{ijk}, Y_{i'jk'}) = var(\beta_j) = \sigma_\beta^2$$

Case 3: Same Factor A and B, Different Replication
If both factor levels are the same (i, j fixed), but different replication ($k \neq k'$):

$$cov(Y_{ijk}, Y_{ijk'}) = var(\alpha_i) + var(\beta_j) + var((\alpha\beta)_{ij}) = \sigma_\alpha^2 + \sigma_\beta^2 + \sigma_{\alpha\beta}^2$$

Table 5.9 Summary of variance-covariance structure

Case	Condition	Covariance
Same factor A, different factor B	i same, $j \neq j'$	σ_α^2
Same factor B, different factor A	j same, $i \neq i'$	σ_β^2
Same factor levels, different replications	i same, j same, $k \neq k'$	$\sigma_\alpha^2 + \sigma_\beta^2 + \sigma_{\alpha\beta}^2$
Different factor levels	$i \neq i', j \neq j'$	0

Case 4: Completely Different Factor Levels
If neither factor A nor B is the same ($i \neq i'$, $j \neq j'$), then:

$$cov(Y_{ijk}, Y_{i'j'k'}) = 0$$

since all random effects are independent across different factor levels.

Summary of Variance-Covariance Structure
Table 5.9.

5.1.5 Two-Way Mixed Effects ANOVA

In a Two-Way Mixed Effects Model, one factor is fixed, while the other is random.
This is often referred to as a mixed effects model or simply a mixed model.

5.1.5.1 Balanced

For a balanced design, the restricted mixed model is:

$$Y_{ijk} = \mu_{..} + \alpha_i + \beta_j + (\alpha\beta)_{ij} + \epsilon_{ijk}$$

where:

- $\mu_{..}$: Overall mean (constant).
- α_i: Fixed effects for Factor A, subject to the constraint $\sum \alpha_i = 0$.
- $\beta_j \sim N(0, \sigma_\beta^2)$ (random effects for Factor B).
- $(\alpha\beta)_{ij} \sim N(0, \frac{a-1}{a}\sigma_{\alpha\beta}^2)$
(interaction effects, constrained so that $\sum_i (\alpha\beta)_{ij} = 0$ for all j). The variance is written as the proportion for convenience, it makes the expected mean squares simpler.
- $\epsilon_{ijk} \sim N(0, \sigma^2)$ (random error).
- $\beta_j, (\alpha\beta)_{ij}, \epsilon_{ijk}$ are pairwise independent.

The restriction on interaction variance ($\frac{a-1}{a}\sigma_{\alpha\beta}^2$) simplifies the expected mean squares, though some sources assume $var((\alpha\beta)_{ij}) = \sigma_{\alpha\beta}^2$.

An unrestricted version of the model removes constraints on interaction terms.

Define:

$$\beta_j = \beta_j^* + (\bar{\alpha\beta})_{ij}^*$$

$$(\alpha\beta)_{ij} = (\alpha\beta)_{ij}^* - (\bar{\alpha\beta})_{ij}^*$$

where β^* and $(\alpha\beta)_{ij}^*$ are unrestricted random effects.

Some consider the restricted model more general, but we use the restricted form for simplicity.

Taking expectations:

$$E(Y_{ijk}) = \mu_{..} + \alpha_i$$

The total variance of responses:

$$var(Y_{ijk}) = \sigma_\beta^2 + \frac{a-1}{a}\sigma_{\alpha\beta}^2 + \sigma^2$$

Covariance Structure

Observations sharing the same **random factor (B)** level are correlated (Table 5.10).

Thus, observations only become independent when they do not share the same random effect.

An advantage of the **restricted mixed model** is that two observations from the same random factor (B) level can be positively or negatively correlated. In the **unrestricted** model, they can only be positively correlated (Table 5.11).

While SS and df are identical across models, the expected mean squares differ, affecting test statistics.

5.1.5.1.1 Hypothesis Testing in Mixed ANOVA

In random ANOVA, we test:

$$H_0 : \sigma^2 = 0 \quad vs. \quad H_a : \sigma^2 > 0$$

Table 5.10 Covariances for different cases

Condition	Covariance
Same i, j, different replications ($k \neq k'$)	$cov(Y_{ijk}, Y_{ijk'}) = \sigma_\beta^2 + \frac{a-1}{a}\sigma_{\alpha\beta}^2$
Same j, different i ($i \neq i'$)	$cov(Y_{ijk}, Y_{i'jk'}) = \sigma_\beta^2 - \frac{1}{a}\sigma_{\alpha\beta}^2$
Different i and j ($i \neq i'$, $j \neq j'$)	$cov(Y_{ijk}, Y_{i'j'k'}) = 0$

Table 5.11 Comparison of fixed, random, and mixed effects models

Mean square	Fixed ANOVA (A, B fixed)	Random ANOVA (A, B random)	Mixed ANOVA (A fixed, B random)
MSA	$\sigma^2 + nb\frac{\sum \alpha_i^2}{a-1}$	$\sigma^2 + nb\sigma_\alpha^2$	$\sigma^2 + nb\frac{\sum_{i=1}^{a} \alpha_i^2}{a-1} + n\sigma_{\alpha\beta}^2$
MSB	$\sigma^2 + na\frac{\sum \beta_j^2}{b-1}$	$\sigma^2 + na\sigma_\beta^2$	$\sigma^2 + na\sigma_\beta^2 + n\sigma_{\alpha\beta}^2$
MSAB	$\sigma^2 + n\frac{\sum (\alpha\beta)_{ij}^2}{(a-1)(b-1)}$	$\sigma^2 + n\sigma_{\alpha\beta}^2$	$\sigma^2 + n\sigma_{\alpha\beta}^2$
MSE	σ^2	σ^2	σ^2

Table 5.12 F-test denominators for factorial ANOVA under different model assumptions

Test for effect of	Fixed ANOVA (A, B fixed)	Random ANOVA (A, B random)	Mixed ANOVA (A fixed, random)
Factor A	$\frac{MSA}{MSE}$	$\frac{MSA}{MSAB}$	$\frac{MSA}{MSAB}$
Factor B	$\frac{MSB}{MSE}$	$\frac{MSB}{MSAB}$	$\frac{MSB}{MSE}$
Interaction (A × B)	$\frac{MSAB}{MSE}$	$\frac{MSAB}{MSE}$	$\frac{MSAB}{MSE}$

using:

$$F = \frac{MSA}{MSAB} \sim F_{a-1,(a-1)(b-1)}$$

For mixed models, the same test statistic is used for:

$$H_0 : \alpha_i = 0, \quad \forall i$$

However, for fixed effects models, the test statistic differs (Table 5.12).

5.1.5.1.2 Variance Component Estimation

In random and mixed effects models, we are interested in estimating variance components.

To estimate σ_β^2:

$$E(\sigma_\beta^2) = \frac{E(MSB) - E(MSE)}{na} = \frac{\sigma^2 + na\sigma_\beta^2 - \sigma^2}{na} = \sigma_\beta^2$$

which is estimated by:

$$\hat{\sigma}_\beta^2 = \frac{MSB - MSE}{na}$$

Confidence intervals for variance components can be approximated using:

- Satterthwaite procedure
- Modified large-sample (MLS) method

5.1.5.1.3 Estimating Fixed Effects in Mixed Models

Fixed effects α_i are estimated by:

$$\hat{\alpha}_i = \bar{Y}_{i..} - \bar{Y}_{...}$$

$$\hat{\mu}_{i.} = \bar{Y}_{...} + (\bar{Y}_{i..} - \bar{Y}_{...}) = \bar{Y}_{i..}$$

Their variances:

$$\sigma^2(\hat{\alpha}_i) = \frac{\sigma^2 + n\sigma^2_{\alpha\beta}}{bn} = \frac{E(MSAB)}{bn}$$

$$s^2(\hat{\alpha}_i) = \frac{MSAB}{bn}$$

5.1.5.1.4 Contrasts on Fixed Effects

For a contrast:

$$L = \sum c_i \alpha_i, \quad \text{where } \sum c_i = 0$$

Estimate:

$$\hat{L} = \sum c_i \hat{\alpha}_i$$

Variance:

$$\sigma^2(\hat{L}) = \sum c_i^2 \sigma^2(\hat{\alpha}_i), \quad s^2(\hat{L}) = \frac{MSAB}{bn} \sum c_i^2$$

5.1.5.2 Unbalanced Two-Way Mixed Effects ANOVA

In an **unbalanced** two-way mixed model (e.g., $a = 2, b = 4$), the model remains:

$$Y_{ijk} = \mu_{..} + \alpha_i + \beta_j + (\alpha\beta)_{ij} + \epsilon_{ijk}$$

where:

- α_i: Fixed effects for Factor A
- $\beta_j \sim N(0, \sigma_\beta^2)$: Random effects for Factor B
- $(\alpha\beta)_{ij} \sim N(0, \frac{\sigma_{\alpha\beta}^2}{2})$: Interaction effects
- $\epsilon_{ijk} \sim N(0, \sigma^2)$: Residual error

5.1.5.2.1 Variance Components

The variance components are:

$$var(\beta_j) = \sigma_\beta^2$$

$$var((\alpha\beta)_{ij}) = \frac{2-1}{2}\sigma_{\alpha\beta}^2 = \frac{\sigma_{\alpha\beta}^2}{2}$$

$$var(\epsilon_{ijk}) = \sigma^2$$

5.1.5.2.2 Expectation and Variance

Taking expectations:

$$E(Y_{ijk}) = \mu_{..} + \alpha_i$$

Total variance:

$$var(Y_{ijk}) = \sigma_\beta^2 + \frac{\sigma_{\alpha\beta}^2}{2} + \sigma^2$$

5.1.5.2.3 Covariance Structure

Observations sharing **Factor B (random effect)** are correlated (Table 5.13).

Thus, only observations within the same random factor level share dependence.

Table 5.13 Covariances for different cases

Condition	Covariance
Same i, j, different replications ($k \neq k'$)	$cov(Y_{ijk}, Y_{ijk'}) = \sigma^2 + \frac{\sigma_{\alpha\beta}^2}{2}$
Same j, different i ($i \neq i'$)	$cov(Y_{ijk}, Y_{i'jk'}) = \sigma_\beta^2 - \frac{\sigma_{\alpha\beta}^2}{2}$
Different i and j ($i \neq i'$, $j \neq j'$)	$cov(Y_{ijk}, Y_{i'j'k'}) = 0$

5.1.5.2.4 Matrix Representation

Assume:

$$\mathbf{Y} \sim N(\mathbf{X}\beta, M)$$

where:

- $\mathbf{X}$: Fixed effects design matrix
- β: Fixed effect coefficients
- M: Block diagonal covariance matrix containing variance components

The density function of $\mathbf{Y}$ is:

$$f(\mathbf{Y}) = \frac{1}{(2\pi)^{N/2}|M|^{1/2}} \exp\left(-\frac{1}{2}(\mathbf{Y} - \mathbf{X}\beta)' M^{-1}(\mathbf{Y} - \mathbf{X}\beta)\right)$$

If variance components were known, we could use Generalized Least Squares:

$$\hat{\beta}_{GLS} = (\mathbf{X}' M^{-1}\mathbf{X})^{-1}\mathbf{X}' M^{-1}\mathbf{Y}$$

However, since variance components $(\sigma^2, \sigma^2_\beta, \sigma^2_{\alpha\beta})$ are unknown, we estimate them using:

- Maximum Likelihood
- Restricted Maximum Likelihood

Maximizing the likelihood:

$$\ln L = -\frac{N}{2}\ln(2\pi) - \frac{1}{2}\ln|M| - \frac{1}{2}(\mathbf{Y} - \mathbf{X}\beta)' M^{-1}(\mathbf{Y} - \mathbf{X}\beta)$$

where:

- $|M|$: Determinant of the variance-covariance matrix
- $(\mathbf{Y} - \mathbf{X}\beta)' M^{-1}(\mathbf{Y} - \mathbf{X}\beta)$: Quadratic form in the likelihood

5.2 Nonparametric ANOVA

When assumptions of **normality** and **equal variance** are not satisfied, we use nonparametric ANOVA tests, which rank the data instead of using raw values.

5.2.1 *Kruskal-Wallis Test (One-Way Nonparametric ANOVA)*

The Kruskal-Wallis test is a generalization of the Wilcoxon rank-sum test to more than two independent samples. It is an alternative to one-way ANOVA when normality is not assumed.

Setup

- $a \geq 2$ independent treatments.
- n_i is the sample size for the i-th treatment.
- Y_{ij} is the j-th observation from the i-th treatment.
- No assumption of normality.
- Assume observations are independent random samples from continuous CDFs $F_1, F_2, \ldots, F_a$.

Hypotheses

$$H_0 : F_1 = F_2 = \cdots = F_a \quad \text{(All distributions are identical)}$$
$$H_a : F_i < F_j \text{ for some } i \neq j$$

If the data come from a location-scale family, the hypothesis simplifies to:

$$H_0 : \theta_1 = \theta_2 = \cdots = \theta_a$$

Procedure

1. Rank all $N = \sum_{i=1}^{a} n_i$ observations in ascending order.
   Let $r_{ij} = rank(Y_{ij})$
   The sum of ranks must satisfy:

$$\sum_i \sum_j r_{ij} = \frac{N(N+1)}{2}$$

2. Compute rank sums and averages:

$$r_{i.} = \sum_{j=1}^{n_i} r_{ij}, \quad \bar{r}_{i.} = \frac{r_{i.}}{n_i}$$

3. Calculate the test statistic:

$$\chi^2_{KW} = \frac{SSTR}{\frac{SSTO}{N-1}}$$

where:

- Treatment Sum of Squares:

$$SSTR = \sum n_i(\bar{r}_{i.} - \bar{r}_{..})^2$$

- Total Sum of Squares:

$$SSTO = \sum_i \sum_j (r_{ij} - \bar{r}_{..})^2$$

- Overall Mean Rank:

$$\bar{r}_{..} = \frac{N+1}{2}$$

4. Compare to a chi-square distribution:
   - For large n_i (≥ 5), $\chi^2_{KW} \sim \chi^2_{a-1}$.
   - Reject H_0 if:

$$\chi^2_{KW} > \chi^2_{(1-\alpha;a-1)}$$

5. Exact Test for Small Samples:
   - Compute all possible rank assignments:

$$\frac{N!}{n_1!n_2!\ldots n_a!}$$

   - Evaluate each Kruskal-Wallis statistic and determine the empirical p-value.

5.2.2 Friedman Test (Nonparametric Two-Way ANOVA)

The Friedman test is a distribution-free alternative to two-way ANOVA when data are measured in a randomized complete block design and normality cannot be assumed.

Setup

- Y_{ij} represents responses from n blocks and r treatments.
- Assume no normality or homogeneity of variance.
- Let F_{ij} be the CDF of Y_{ij}, corresponding to observed values.

Hypotheses

$$H_0 : F_{i1} = F_{i2} = \cdots = F_{ir} \quad \forall i \quad \text{(Identical distributions within each block)}$$
$$H_a : F_{ij} < F_{ij'} \text{ for some } j \neq j' \quad \forall i$$

For location-scale families, the hypothesis simplifies to:

$$H_0 : \tau_1 = \tau_2 = \cdots = \tau_r$$
$$H_a : \tau_j > \tau_{j'} \text{ for some } j \neq j'$$

Procedure

1. Rank observations within each block separately (ascending order).
   - If there are ties, assign average ranks.
2. Compute test statistic:

$$\chi_F^2 = \frac{SSTR}{\frac{SSTR+SSE}{n(r-1)}}$$

where:

- Treatment Sum of Squares:

$$SSTR = n\sum(\bar{r}_{.j} - \bar{r}_{..})^2$$

- Error Sum of Squares:

$$SSE = \sum_i \sum_j (r_{ij} - \bar{r}_{.j})^2$$

- Mean Ranks:

$$\bar{r}_{.j} = \frac{\sum_i r_{ij}}{n}, \quad \bar{r}_{..} = \frac{r+1}{2}$$

3. Alternative Formula for Large Samples (No Ties):
   If no ties, Friedman's statistic simplifies to:

$$\chi_F^2 = \left[\frac{12}{nr(n+1)}\sum_j r_{.j}^2\right] - 3n(r+1)$$

4. Compare to a chi-square distribution:
   - For large n, $\chi_F^2 \sim \chi_{r-1}^2$.
   - Reject H_0 if:

$$\chi_F^2 > \chi_{(1-\alpha;r-1)}^2$$

5. Exact Test for Small Samples:
   - Compute all possible ranking permutations:

$$(r!)^n$$

   - Evaluate each Friedman statistic and determine the empirical p-value.

5.3 Randomized Block Designs

To improve the precision of treatment comparisons, we can **reduce variability** among experimental units by grouping them into **blocks**.
Each block contains **homogeneous units**, reducing the impact of nuisance variation.

Key Principles of Blocking
- Within each block, treatments are randomly assigned to units.
- The number of units per block is a multiple of the number of factor combinations.
- Commonly, each treatment appears once per block.

Benefits of Blocking
- Reduction in variability of treatment effect estimates
  - Improved power for t-tests and F-tests
  - Narrower confidence intervals
  - Smaller mean square error (MSE)
- Allows comparison of treatments across different conditions (captured by blocks)

Potential Downsides of Blocking
- If blocks are not chosen well, degrees of freedom are wasted on negligible block effects.
- This reduces df for t-tests and F-tests without reducing MSE, causing a small loss of power.

5.3.1 Random Block Effects with Additive Effects

The statistical model for a randomized block design:

$$Y_{ij} = \mu_{..} + \rho_i + \tau_j + \epsilon_{ij}$$

where:

- $i = 1, 2, \ldots, n$ (Blocks)
- $j = 1, 2, \ldots, r$ (Treatments)
- $\mu_{..}$: Overall mean response (averaged across all blocks and treatments)
- ρ_i: Block effect (average difference for the i-th block), constrained such that:

$$\sum_i \rho_i = 0$$

- τ_j: Treatment effect (average across blocks), constrained such that:

$$\sum_j \tau_j = 0$$

- $\epsilon_{ij} \sim iidN(0, \sigma^2)$: Random experimental error

Interpretation of the Model

- Block and treatment effects are additive.
- The difference in average response between any two treatments is the same within each block:

$$(\mu_{..} + \rho_i + \tau_j) - (\mu_{..} + \rho_i + \tau_j') = \tau_j - \tau_j'$$

- This ensures that blocking only affects variability, not treatment comparisons.

Estimators of Model Parameters

- **Overall Mean**:

$$\hat{\mu} = \bar{Y}_{..}$$

- **Block Effects**:

$$\hat{\rho}_i = \bar{Y}_{i.} - \bar{Y}_{..}$$

- **Treatment Effects**:

$$\hat{\tau}_j = \bar{Y}_{.j} - \bar{Y}_{..}$$

Table 5.14 ANOVA table for randomized block design

Source of variation	Sum of squares (SS)	Degrees of freedom (df)	Fixed treatments (E(MS))	Random treatments (E(MS))
Blocks	$r\sum_i(\bar{Y}_{i.}-\bar{Y}_{..})^2$	$n-1$	$\sigma^2+r\frac{\sum\rho_i^2}{n-1}$	$\sigma^2+r\frac{\sum\rho_i^2}{n-1}$
Treatments	$n\sum_j(\bar{Y}_{.j}-\bar{Y}_{..})^2$	$r-1$	$\sigma^2+n\frac{\sum\tau_j^2}{r-1}$	$\sigma^2+n\sigma_\tau^2$
Error	$\sum_i\sum_j(Y_{ij}-\bar{Y}_{i.}-\bar{Y}_{.j}+\bar{Y}_{..})^2$	$(n-1)(r-1)$	σ^2	σ^2
Total	$SSTO$	$nr-1$	-	-

- **Fitted Response:**

$$\hat{Y}_{ij}=\bar{Y}_{..}+(\bar{Y}_{i.}-\bar{Y}_{..})+(\bar{Y}_{.j}-\bar{Y}_{..})$$

  Simplifies to:

$$\hat{Y}_{ij}=\bar{Y}_{i.}+\bar{Y}_{.j}-\bar{Y}_{..}$$

- **Residuals**:

$$e_{ij}=Y_{ij}-\hat{Y}_{ij}=Y_{ij}-\bar{Y}_{i.}-\bar{Y}_{.j}+\bar{Y}_{..}$$

The ANOVA decomposition partitions total variability into contributions from blocks, treatments, and residual error (Table 5.14).

5.3.2 F-Tests in Randomized Block Designs

To test for treatment effects, we use an F-test:

For **fixed treatment effects**:

$$H_0:\tau_1=\tau_2=\cdots=\tau_r=0 \quad \text{(No treatment effect)}$$

$$H_a:\text{Not all } \tau_j=0$$

For **random treatment effects**:

$$H_0:\sigma_\tau^2=0 \quad \text{(No variance in treatment effects)}$$

$$H_a:\sigma_\tau^2\neq 0$$

In both cases, the test statistic is:

$$F = \frac{MSTR}{MSE}$$

Reject H_0 if:

$$F > f_{(1-\alpha;r-1,(n-1)(r-1))}$$

Why Not Use an F-Test for Blocks?
We do not test for block effects because:

1. Blocks are assumed to be different a priori.
2. Randomization occurs within each block, ensuring treatments are comparable.

Efficiency Gain from Blocking
To measure the improvement in precision, compare the mean square error (MSE) in a completely randomized design vs. a randomized block design.

Estimated variance in a CRD:

$$\hat{\sigma}^2_{CR} = \frac{(n-1)MSBL + n(r-1)MSE}{nr-1}$$

Estimated variance in an RBD:

$$\hat{\sigma}^2_{RB} = MSE$$

Relative efficiency:

$$\frac{\hat{\sigma}^2_{CR}}{\hat{\sigma}^2_{RB}}$$

- If greater than 1, blocking reduces experimental error.
- The percentage reduction in required sample size for an RBD:

$$\left(\frac{\hat{\sigma}^2_{CR}}{\hat{\sigma}^2_{RB}} - 1\right) \times 100\%$$

Random Blocks and Mixed Models
If blocks are randomly selected, they are treated as random effects.

That is, if the experiment were repeated, a new set of blocks would be selected, with:

$$\rho_1, \rho_2, \ldots, \rho_i \sim N(0, \sigma^2_\rho)$$

The model remains:

$$Y_{ij} = \mu_{..} + \rho_i + \tau_j + \epsilon_{ij}$$

where:

- $\mu_{..}$ is fixed.
- $\rho_i \sim iidN(0, \sigma_\rho^2)$ (random block effects).
- τ_j is fixed (or random, with $\sum \tau_j = 0$).
- $\epsilon_{ij} \sim iidN(0, \sigma^2)$.

5.3.3 Variance and Covariance Structure

For **fixed treatment effects**:

$$E(Y_{ij}) = \mu_{..} + \tau_j$$
$$var(Y_{ij}) = \sigma_\rho^2 + \sigma^2$$

Observations **within the same block** are **correlated**:

$$cov(Y_{ij}, Y_{ij'}) = \sigma_\rho^2, \quad j \neq j'$$

Observations **from different blocks** are **independent**:

$$cov(Y_{ij}, Y_{i'j'}) = 0, \quad i \neq i', j \neq j'$$

The **intra-block correlation** (Table 5.15):

$$\frac{\sigma_\rho^2}{\sigma^2 + \sigma_\rho^2}$$

Table 5.15 Expected mean squares for fixed treatments

Source	SS	E(MS)
Blocks	SSBL	$\sigma^2 + r\sigma_\rho^2$
Treatments	SSTR	$\sigma^2 + n\frac{\sum \tau_j^2}{r-1}$
Error	SSE	σ^2

5.3.4 Random Block Effects with Interaction

When block-treatment interaction exists, we modify the model:

$$Y_{ij} = \mu_{..} + \rho_i + \tau_j + (\rho\tau)_{ij} + \epsilon_{ij}$$

where:

- $\rho_i \sim iidN(0, \sigma_\rho^2)$ (random).
- τ_j is fixed ($\sum \tau_j = 0$).
- $(\rho\tau)_{ij} \sim N(0, \frac{r-1}{r}\sigma_{\rho\tau}^2)$, constrained such that:

$$\sum_j (\rho\tau)_{ij} = 0, \quad \forall i$$

- Covariance between interaction terms:

$$cov((\rho\tau)_{ij}, (\rho\tau)_{ij'}) = -\frac{1}{r}\sigma_{\rho\tau}^2, \quad j \neq j'$$

- $\epsilon_{ij} \sim iidN(0, \sigma^2)$.

Variance and Covariance with Interaction

- **Expectation**:

$$E(Y_{ij}) = \mu_{..} + \tau_j$$

- **Total variance**:

$$var(Y_{ij}) = \sigma_\rho^2 + \frac{r-1}{r}\sigma_{\rho\tau}^2 + \sigma^2$$

- **Within-block covariance**:

$$cov(Y_{ij}, Y_{ij'}) = \sigma_\rho^2 - \frac{1}{r}\sigma_{\rho\tau}^2, \quad j \neq j'$$

- **Between-block covariance**:

$$cov(Y_{ij}, Y_{i'j'}) = 0, \quad i \neq i', j \neq j'$$

The sum of squares and degrees of freedom for interaction model are the same as those for the additive model. The difference exists only in the expected mean squares (Table 5.16).

Table 5.16 ANOVA table with interaction effects

Source	SS	df	E(MS)
Blocks	$SSBL$	$n-1$	$\sigma^2 + r\sigma_\rho^2$
Treatments	$SSTR$	$r-1$	$\sigma^2 + \sigma_{\rho\tau}^2 + n\frac{\sum\tau_j^2}{r-1}$
Error	SSE	$(n-1)(r-1)$	$\sigma^2 + \sigma_{\rho\tau}^2$

- No exact test is possible for block effects when interaction is present. (Not important if blocks are used primarily to reduce experimental error variability.)
- $E(MSE) = \sigma^2 + \sigma_{\rho\tau}^2$ the error term variance and interaction variance $\sigma_{\rho\tau}^2$. We can't estimate these components separately with this model. The two are **confounded**.
- If more than one observation per treatment block combination, one can consider interaction with fixed block effects, which is called **generalized randomized block designs** (multifactor analysis).

5.3.5 *Tukey Test of Additivity*

Tukey's 1-degree-of-freedom test for additivity provides a formal test for interaction effects between blocks and treatments in a randomized block design.

This test can also be used in two-way ANOVA when there is only one observation per cell.

In a randomized block design, an additive model assumes:

$$Y_{ij} = \mu_{..} + \rho_i + \tau_j + \epsilon_{ij}$$

where:

- $\mu_{..}$ = overall mean
- ρ_i = block effect
- τ_j = treatment effect
- ϵ_{ij} = random error, $iidN(0, \sigma^2)$

To test for interaction, we introduce a less restricted interaction term:

$$(\rho\tau)_{ij} = D\rho_i\tau_j$$

where D is a constant measuring interaction strength.

Thus, the interaction model becomes:

$$Y_{ij} = \mu_{..} + \rho_i + \tau_j + D\rho_i\tau_j + \epsilon_{ij}$$

The least squares estimate (or MLE) of D is:

$$\hat{D} = \frac{\sum_i \sum_j \rho_i \tau_j Y_{ij}}{\sum_i \rho_i^2 \sum_j \tau_j^2}$$

Replacing ρ_i and τ_j with their estimates:

$$\hat{D} = \frac{\sum_i \sum_j (\bar{Y}_{i.} - \bar{Y}_{..})(\bar{Y}_{.j} - \bar{Y}_{..}) Y_{ij}}{\sum_i (\bar{Y}_{i.} - \bar{Y}_{..})^2 \sum_j (\bar{Y}_{.j} - \bar{Y}_{..})^2}$$

The sum of squares for interaction is:

$$SS_{int} = \sum_i \sum_j \hat{D}^2 (\bar{Y}_{i.} - \bar{Y}_{..})^2 (\bar{Y}_{.j} - \bar{Y}_{..})^2$$

ANOVA Decomposition
The total sum of squares (SSTO) is decomposed as:

$$SSTO = SSBL + SSTR + SS_{int} + SS_{Rem}$$

where:

- $SSBL$ = Sum of squares due to blocks
- $SSTR$ = Sum of squares due to treatments
- SS_{int} = Interaction sum of squares
- SS_{Rem} = Remainder sum of squares, computed as:

$$SS_{Rem} = SSTO - SSBL - SSTR - SS_{int}$$

We test:

$$H_0 : D = 0 \quad \text{(No interaction present)}$$
$$H_a : D \neq 0 \quad \text{(Interaction of form } D\rho_i\tau_j \text{ present)}$$

If $D = 0$, then SS_{int} and SS_{Rem} are independent and follow:

$$SS_{int} \sim \chi^2_1, \quad SS_{Rem} \sim \chi^2_{(rn-r-n)}$$

Thus, the F-statistic for testing interaction is:

$$F = \frac{SS_{int}/1}{SS_{Rem}/(rn - r - n)}$$

which follows an F-distribution:

$$F \sim F_{(1,nr-r-n)}$$

We reject H_0 if:

$$F > f_{(1-\alpha;1,nr-r-n)}$$

5.4 Nested Designs

A nested design occurs when one factor is entirely contained within another. This differs from a crossed design, where all levels of one factor are present across all levels of another factor.

- **Crossed Design**: If Factor B is crossed with Factor A, then each level of Factor B appears at every level of Factor A.
- **Nested Design**: If Factor B is nested within Factor A, then each level of Factor B is unique to a particular level of Factor A.

Thus, if Factor B is nested within Factor A:

- Level 1 of B within A = 1 has nothing in common with
- Level 1 of B within A = 2.

Types of Factors

- **Classification Factors**: Factors that **cannot be manipulated** (e.g., geographical regions, subjects).
- **Experimental Factors**: Factors that are **randomly assigned** in an experiment.

5.4.1 Two-Factor Nested Design

We consider a nested two-factor model where:

- Factor A has a levels.
- Factor B is nested within Factor A, with b levels per level of A.
- Both factors are fixed.
- All treatment means are equally important.

The mean response at level i of Factor A:

$$\mu_{i.} = \frac{1}{b} \sum_j \mu_{ij}$$

The main effect of Factor A:

$$\alpha_i = \mu_{i.} - \mu_{..}$$

where:

$$\mu_{..} = \frac{1}{ab}\sum_i \sum_j \mu_{ij} = \frac{1}{a}\sum_i \mu_{i.}$$

with the constraint:

$$\sum_i \alpha_i = 0$$

The nested effect of Factor B within A is denoted as $\beta_{j(i)}$, where:

$$\begin{aligned}\beta_{j(i)} &= \mu_{ij} - \mu_{i.} \\ &= \mu_{ij} - \alpha_i - \mu_{..}\end{aligned}$$

with the restriction:

$$\sum_j \beta_{j(i)} = 0, \quad \forall i = 1, \ldots, a$$

Since $\beta_{j(i)}$ is the specific effect of the j-th level of factor B nested within the i-th level of factor A, the full model can be written as:

$$\mu_{ij} = \mu_{..} + \alpha_i + \beta_{j(i)}$$

or equivalently:

$$\mu_{ij} = \mu_{..} + (\mu_{i.} - \mu_{..}) + (\mu_{ij} - \mu_{i.})$$

The statistical model for a two-factor nested design is:

$$Y_{ijk} = \mu_{..} + \alpha_i + \beta_{j(i)} + \epsilon_{ijk}$$

where:

- Y_{ijk} = response for the k-th observation when:
  - **Factor A** is at level i.
  - **Factor B** (nested within A) is at level j.
- $\mu_{..}$ = overall mean.
- α_i = main effect of **Factor A** (subject to: $\sum_i \alpha_i = 0$).

Table 5.17 Estimators for parameters in nested designs

Parameter	Estimator
$\mu_{..}$	$\bar{Y}_{...}$
α_i	$\bar{Y}_{i..} - \bar{Y}_{...}$
$\beta_{j(i)}$	$\bar{Y}_{ij.} - \bar{Y}_{i..}$
$\hat{Y}_{ijk}$	$\bar{Y}_{ij.}$

Table 5.18 ANOVA table for nested designs

Source of variation	SS	df	MS	E(MS)
Factor A	SSA	$a-1$	MSA	$\sigma^2 + bn\frac{\sum \alpha_i^2}{a-1}$
Factor B (A)	$SSB(A)$	$a(b-1)$	$MSB(A)$	$\sigma^2 + n\frac{\sum \beta_{j(i)}^2}{a(b-1)}$
Error	SSE	$ab(n-1)$	MSE	σ^2
Total	$SSTO$	$abn-1$		

- $\beta_{j(i)}$ = nested effect of **Factor B within A** (subject to: $\sum_j \beta_{j(i)} = 0$ for all i).
- $\epsilon_{ijk} \sim iidN(0, \sigma^2)$ = random error.

  Thus, the expected value and variance are:

$$E(Y_{ijk}) = \mu_{..} + \alpha_i + \beta_{j(i)}$$
$$var(Y_{ijk}) = \sigma^2$$

Note: There is no interaction term in a nested model, because Factor B levels are unique within each level of A.

The least squares and maximum likelihood estimates (Table 5.17):
The residual error:

$$e_{ijk} = Y_{ijk} - \bar{Y}_{ij.}$$

The total sum of squares (SSTO) is partitioned as (Table 5.18):

$$SSTO = SSA + SSB(A) + SSE$$

where:

$$\sum_i \sum_j \sum_k (Y_{ijk} - \bar{Y}_{...})^2 = bn \sum_i (\bar{Y}_{i..} - \bar{Y}_{...})^2 + n \sum_i \sum_j (\bar{Y}_{ij.} - \bar{Y}_{i..})^2 + \sum_i \sum_j \sum_k (Y_{ijk} - \bar{Y}_{ij.})^2$$

5.4.1.1 Tests for Factor Effects

- Factor A:

$$F = \frac{MSA}{MSB(A)} \sim F_{(a-1,a(b-1))}$$

Reject H_0 if $F > f_{(1-\alpha;a-1,a(b-1))}$.
- Factor B within A:

$$F = \frac{MSB(A)}{MSE} \sim F_{(a(b-1),ab(n-1))}$$

Reject H_0 if $F > f_{(1-\alpha;a(b-1),ab(n-1))}$.

5.4.1.2 Testing Factor Effect Contrasts

A contrast is a linear combination of factor level means:

$$L = \sum c_i \mu_i, \quad \text{where} \quad \sum c_i = 0$$

The estimated contrast:

$$\hat{L} = \sum c_i \bar{Y}_{i..}$$

The confidence interval for L:

$$\hat{L} \pm t_{(1-\alpha/2;df)} s(\hat{L})$$

where:

$$s^2(\hat{L}) = \sum c_i^2 s^2(\bar{Y}_{i..}), \quad \text{where} \quad s^2(\bar{Y}_{i..}) = \frac{MSE}{bn}, \quad df = ab(n-1)$$

5.4.1.3 Testing Treatment Means

For treatment means, a similar approach applies:

$$L = \sum c_i \mu_{.j}, \quad \hat{L} = \sum c_i \bar{Y}_{ij}$$

The confidence limits for L:

$$\hat{L} \pm t_{(1-\alpha/2;(n-1)ab)} s(\hat{L})$$

where:

$$s^2(\hat{L}) = \frac{MSE}{n} \sum c_i^2$$

5.4.2 Unbalanced Nested Two-Factor Designs

When Factor B has different levels for different levels of Factor A, the design is unbalanced.

$$Y_{ijk} = \mu_{..} + \alpha_i + \beta_{j(i)} + \epsilon_{ijk}$$

$$\sum_{i=1}^{2} \alpha_i = 0, \quad \sum_{j=1}^{3} \beta_{j(1)} = 0, \quad \sum_{j=1}^{2} \beta_{j(2)} = 0$$

where:

- Factor A: $i = 1, 2$.
- Factor B (nested in A): $j = 1, \ldots, b_i$.
- Observations: $k = 1, \ldots, n_{ij}$.

  Example case:

- $b_1 = 3, b_2 = 2$ (Factor B has different levels for A).
- $n_{11} = n_{13} = 2, n_{12} = 1, n_{21} = n_{22} = 2$.
- Parameters: $\alpha_1, \beta_{1(1)}, \beta_{2(1)}, \beta_{1(2)}$.

  Constraints:

$$\alpha_2 = -\alpha_1, \quad \beta_{3(1)} = -\beta_{1(1)} - \beta_{2(1)}, \quad \beta_{2(2)} = -\beta_{1(2)}$$

The unbalanced design can be modeled using **indicator variables**:

1. **Factor A (School Level):**

$$X_1 = \begin{cases} 1 & \text{if observation from school 1} \\ -1 & \text{if observation from school 2} \end{cases}$$

2. **Factor B (Instructor Within School 1):**

$$X_2 = \begin{cases} 1 & \text{if observation from instructor 1 in school 1} \\ -1 & \text{if observation from instructor 3 in school 1} \\ 0 & \text{otherwise} \end{cases}$$

3. **Factor B (Instructor Within School 1):**

$$X_3 = \begin{cases} 1 & \text{if observation from instructor 2 in school 1} \\ -1 & \text{if observation from instructor 3 in school 1} \\ 0 & \text{otherwise} \end{cases}$$

4. **Factor B (Instructor Within School 1):**

$$X_4 = \begin{cases} 1 & \text{if observation from instructor 1 in school 1} \\ -1 & \text{if observation from instructor 2 in school 1} \\ 0 & \text{otherwise} \end{cases}$$

Using these indicator variables, the **full regression model** is:

$$Y_{ijk} = \mu_{..} + \alpha_1 X_{ijk1} + \beta_{1(1)} X_{ijk2} + \beta_{2(1)} X_{ijk3} + \beta_{1(2)} X_{ijk4} + \epsilon_{ijk}$$

where X_1, X_2, X_3, X_4 represent different factor effects.

5.4.3 *Random Factor Effects*

If factors are **random** (Tables 5.19 and 5.20):

$$\alpha_1 \sim iidN(0, \sigma_\alpha^2)$$
$$\beta_{j(i)} \sim iidN(0, \sigma_\beta^2)$$

Another way to increase precision in treatment comparisons is by adjusting for covariates using regression models. This is called Analysis of Covariance (ANCOVA).

Why Use ANCOVA?

- Reduces variability by accounting for covariate effects.

Table 5.19 Expected mean squares for random effects

Mean square	Expected mean squares (A fixed, B random)	Expected mean squares (A random, B random)
MSA	$\sigma^2 + n\sigma_\beta^2 + bn\frac{\sum \alpha_i^2}{a-1}$	$\sigma^2 + bn\sigma_\alpha^2 + n\sigma_\beta^2$
MSB(A)	$\sigma^2 + n\sigma_\beta^2$	$\sigma^2 + n\sigma_\beta^2$
MSE	σ^2	σ^2

Table 5.20 F-tests for factor effects

Factor	F-test (A fixed, B random)	F-test (A random, B random)
Factor A	$\frac{MSA}{MSB(A)}$	$\frac{MSA}{MSB(A)}$
Factor B(A)	$\frac{MSB(A)}{MSE}$	$\frac{MSB(A)}{MSE}$

- Increases statistical power by removing nuisance variation.
- Combines ANOVA and regression for more precise comparisons.

5.5 Sample Size Planning for ANOVA

5.5.1 *Balanced Designs*

Choosing an appropriate sample size for an ANOVA study requires ensuring sufficient power while balancing practical constraints.

5.5.2 *Single-Factor Studies*

5.5.2.1 Fixed Cell Means Model

The probability of rejecting H_0 when it is false (power) is given by:

$$P(F > f_{(1-\alpha;a-1,N-a)}|\phi) = 1 - \beta$$

where:

- ϕ is the non-centrality parameter (measuring the inequality among treatment means μ_i):

$$\phi = \frac{1}{\sigma}\sqrt{\frac{n}{a}\sum_i(\mu_i - \mu_.)^2}, \quad (n_i \equiv n)$$

- $\mu_.$ is the overall mean:

$$\mu_. = \frac{\sum \mu_i}{a}$$

To determine power, we use the noncentral F distribution.

Using Power Tables
Power tables can be used directly when:

1. The effects are fixed.
2. The design is balanced.
3. The minimum range of factor level means Δ is known:

$$\Delta = \max(\mu_i) - \min(\mu_i)$$

Thus, the required inputs are:

- Significance level (α)
- Minimum range of means (Δ)
- Error standard deviation (σ)
- Power ($1 - \beta$)

Notes on Sample Size Sensitivity

- When Δ/σ is small, sample size requirements increase dramatically.
- Lowering α or β increases required sample sizes.
- Errors in estimating σ can significantly impact sample size calculations.

5.5.3 *Multifactor Studies*

The same noncentral F tables apply for multifactor models.

5.5.3.1 Two-Factor Fixed Effects Model

5.5.3.1.1 Test for Interaction Effects

The non-centrality parameter:

$$\phi = \frac{1}{\sigma}\sqrt{\frac{n\sum_i\sum_j(\alpha\beta)_{ij}^2}{(a-1)(b-1)+1}}$$

or equivalently:

$$\phi = \frac{1}{\sigma}\sqrt{\frac{n\sum_i\sum_j(\mu_{ij} - \mu_{i.} - \mu_{.j} + \mu_{..})^2}{(a-1)(b-1)+1}}$$

where degrees of freedom are:

$$\upsilon_1 = (a-1)(b-1)$$
$$\upsilon_2 = ab(n-1)$$

5.5.3.1.2 Test for Factor *A* Main Effects

The non-centrality parameter:

$$\phi = \frac{1}{\sigma}\sqrt{\frac{nb\sum\alpha_i^2}{a}}$$

or equivalently:

$$\phi = \frac{1}{\sigma}\sqrt{\frac{nb\sum(\mu_{i.}-\mu_{..})^2}{a}}$$

where degrees of freedom are:

$$\upsilon_1 = a-1$$
$$\upsilon_2 = ab(n-1)$$

5.5.3.1.3 Test for Factor *B* Main Effects

The non-centrality parameter:

$$\phi = \frac{1}{\sigma}\sqrt{\frac{na\sum\beta_j^2}{b}}$$

or equivalently:

$$\phi = \frac{1}{\sigma}\sqrt{\frac{na\sum(\mu_{.j}-\mu_{..})^2}{b}}$$

where degrees of freedom are:

$$\upsilon_1 = b-1$$
$$\upsilon_2 = ab(n-1)$$

5.5.4 *Procedure for Sample Size Selection*

1. Specify the minimum range of Factor A means.
2. Obtain sample size from power tables using $r = a$.
   - The resulting sample size is bn, from which n can be derived.
3. Repeat steps 1–2 for Factor B.
4. Choose the larger sample size from the calculations for Factors A and B.

5.5.5 *Randomized Block Experiments*

Analogous to completely randomized designs . The power of the F-test for treatment effects for randomized block design uses the same non-centrality parameter as completely randomized design:

$$\phi = \frac{1}{\sigma}\sqrt{\frac{n}{r}\sum(\mu_i - \mu_.)^2}$$

However, the power level is different from the randomized block design because

- Error variance σ^2 is different.
- df(MSE) is different.

5.6 Single-Factor Covariance Model

The **single-factor covariance model** (Analysis of Covariance (ANCOVA)) accounts for both treatment effects and a continuous covariate:

$$Y_{ij} = \mu_. + \tau_i + \gamma(X_{ij} - \bar{X}_{..}) + \epsilon_{ij}$$

for $i = 1, \ldots, r$ (treatments) and $j = 1, \ldots, n_i$ (observations per treatment).

- $\mu_.$: Overall mean response
- τ_i: Fixed treatment effects ($\sum \tau_i = 0$)
- γ: Fixed regression coefficient (relationship between covariate X and response Y)
- X_{ij}: Observed covariate (fixed, not random)
- $\epsilon_{ij} \sim iidN(0, \sigma^2)$: Independent random errors

If we use γX_{ij} directly (without centering), then $\mu_.$ is no longer the overall mean. Thus, centering the covariate is necessary to maintain interpretability.

Expectation and Variance

$$E(Y_{ij}) = \mu_. + \tau_i + \gamma(X_{ij} - \bar{X}_{..})$$
$$var(Y_{ij}) = \sigma^2$$

Since $Y_{ij} \sim N(\mu_{ij}, \sigma^2)$, we express:

$$\mu_{ij} = \mu_. + \tau_i + \gamma(X_{ij} - \bar{X}_{..})$$

where $\sum \tau_i = 0$. The mean response μ_{ij} is a regression line with intercept $\mu_. + \tau_i$ and slope γ for each treatment i.

Key Assumptions

1. All treatments share the same slope (γ).
2. No interaction between treatment and covariate (parallel regression lines).
3. If slopes differ, ANCOVA is not appropriate $\rightarrow$ use separate regressions per treatment.

   A more general model allows multiple covariates:

$$Y_{ij} = \mu_. + \tau_i + \gamma_1(X_{ij1} - \bar{X}_{..1}) + \gamma_2(X_{ij2} - \bar{X}_{..2}) + \epsilon_{ij}$$

Using indicator variables for treatments:
For treatment $i = 1$:

$$l_1 = \begin{cases} 1 & \text{if case belongs to treatment 1} \\ -1 & \text{if case belongs to treatment } r \\ 0 & \text{otherwise} \end{cases}$$

For treatment $i = r - 1$:

$$l_{r-1} = \begin{cases} 1 & \text{if case belongs to treatment } r-1 \\ -1 & \text{if case belongs to treatment } r \\ 0 & \text{otherwise} \end{cases}$$

Defining $x_{ij} = X_{ij} - \bar{X}_{..}$, the regression model is:

$$Y_{ij} = \mu_. + \tau_1 l_{ij,1} + \cdots + \tau_{r-1} l_{ij,r-1} + \gamma x_{ij} + \epsilon_{ij}$$

where $I_{ij,1}$ is the indicator variable l_1 for the j-th case in treatment i.

The treatment effects (τ_i) are simply regression coefficients for the indicator variables.

5.6.1 Statistical Inference for Treatment Effects

To test treatment effects:

$$H_0 : \tau_1 = \tau_2 = \cdots = 0$$
$$H_a : \text{Not all } \tau_i = 0$$

1. **Full Model (with treatment effects):**

$$Y_{ij} = \mu_. + \tau_i + \gamma X_{ij} + \epsilon_{ij}$$

2. **Reduced Model (without treatment effects):**

$$Y_{ij} = \mu_. + \gamma X_{ij} + \epsilon_{ij}$$

F-Test for Treatment Effects
The test statistic is:

$$F = \frac{SSE(R) - SSE(F)}{(N-2) - (N-(r+1))} \Big/ \frac{SSE(F)}{N-(r+1)}$$

where:

- $SSE(R)$: Sum of squared errors for the **reduced model**
- $SSE(F)$: Sum of squared errors for the **full model**
- N: Total number of observations
- r: Number of treatment groups

  Under H_0, the statistic follows an F-distribution:

$$F \sim F_{(r-1, N-(r+1))}$$

Comparisons of Treatment Effects
For $r = 3$, we estimate (Table 5.21):

Table 5.21 Estimates and variances for treatment comparisons

Comparison	Estimate	Variance of estimator
$\tau_1 - \tau_2$	$\hat{\tau}_1 - \hat{\tau}_2$	$var(\hat{\tau}_1) + var(\hat{\tau}_2) - 2cov(\hat{\tau}_1, \hat{\tau}_2)$
$\tau_1 - \tau_3$	$2\hat{\tau}_1 + \hat{\tau}_2$	$4var(\hat{\tau}_1) + var(\hat{\tau}_2) - 4cov(\hat{\tau}_1, \hat{\tau}_2)$
$\tau_2 - \tau_3$	$\hat{\tau}_1 + 2\hat{\tau}_2$	$var(\hat{\tau}_1) + 4var(\hat{\tau}_2) - 4cov(\hat{\tau}_1, \hat{\tau}_2)$

5.6.2 Testing for Parallel Slopes

To check if slopes differ across treatments, we use the model:

$$Y_{ij} = \mu_{.} + \tau_1 I_{ij,1} + \tau_2 I_{ij,2} + \gamma X_{ij} + \beta_1 I_{ij,1} X_{ij} + \beta_2 I_{ij,2} X_{ij} + \epsilon_{ij}$$

where:

- β_1, β_2: Interaction coefficients (slope differences across treatments)

Hypothesis Test

$$H_0 : \beta_1 = \beta_2 = 0 \quad \text{(Slopes are equal)}$$
$$H_a : \text{At least one } \beta \neq 0 \quad \text{(Slopes differ)}$$

If the F-test fails to reject H_0, then we assume parallel slopes.

5.6.3 Adjusted Means

The adjusted treatment means account for covariate effects:

$$Y_{i.}(\text{adj}) = \bar{Y}_{i.} - \hat{\gamma}(\bar{X}_{i.} - \bar{X}_{..})$$

where:

- $\bar{Y}_{i.}$: Observed mean response for treatment i
- $\hat{\gamma}$: Estimated regression coefficient
- $\bar{X}_{i.}$: Mean covariate value for treatment i
- $\bar{X}_{..}$: Overall mean covariate value

This provides estimated treatment means after controlling for covariate effects.

This chapter has provided an exploration of ANOVA, a foundational technique for comparing group means across multiple experimental conditions. Beginning with the Completely Randomized Design, we established the basic framework for understanding how ANOVA partitions variability. We then extended the discussion to Nonparametric ANOVA, accommodating situations where the assumptions of traditional ANOVA are violated.

Subsequent sections introduced advanced designs such as Randomized Block and Nested Designs, which offer increased precision and control over variability by accounting for structured sources of heterogeneity. We also addressed Sample Size Planning, emphasizing the importance of statistical power and design efficiency. Finally, we examined the Single-Factor Covariance Model, integrating covariates into the ANOVA framework to adjust for confounding variables and improve estimation accuracy.

Together, these topics equip the reader with the methodological rigor needed to design, implement, and interpret experiments involving multiple group comparisons in complex real-world settings.

Chapter 6
Multivariate Methods

In the previous section on ANOVA, we focused on comparing means across multiple groups under the assumption of a **single response variable**. This framework is powerful and widely used, but it quickly becomes inadequate when dealing with the complexities of real-world business data, where outcomes are rarely isolated.

In many practical settings—particularly in marketing, finance, and operations—analysts are often faced with multiple, interrelated response variables that must be considered together to draw meaningful conclusions. Consider the following scenarios:

- **Marketing Analytics**: When evaluating customer segments, it is insufficient to look at purchase frequency in isolation. Instead, we must consider multiple facets such as purchase frequency, basket size, brand perception, and loyalty metrics. These variables are often correlated, and analyzing them jointly can reveal deeper insights into customer behavior and strategic positioning.
- **Financial Portfolio Analysis**: In portfolio management, assessing the expected return of an asset is only half the picture. Effective risk management requires understanding the covariation of returns across multiple assets, especially when constructing diversified portfolios. Here, the interplay between asset returns becomes central to estimating overall portfolio risk.
- **Operational Strategy**: In supply chain optimization, factors like delivery time, cost efficiency, and service quality are interconnected. Modeling them simultaneously allows managers to balance trade-offs and optimize performance across the system.

Multivariate methods extend classical statistical techniques to accommodate **multiple dependent variables** at once, allowing for a more comprehensive understanding of complex phenomena. These methods not only account for the individual behavior of each variable but also for the relationships among them. At the mathematical core of these techniques is the covariance matrix, a generalization of variance that captures the pairwise linear associations between variables.

M. Nguyen, *Experimental Design*, https://doi.org/10.1007/978-3-032-01839-7_6

Multivariate analysis is not merely a technical extension of univariate methods—it is a conceptual shift. It reflects the reality that variables rarely operate in isolation, and that richer, more accurate conclusions often arise from analyzing their joint behavior.

6.1 Basic Understanding

6.1.1 Multivariate Random Vectors

Let $y_1, \ldots, y_p$ be random variables, possibly correlated, with means $\mu_1, \ldots, \mu_p$. We define the random vector:

$$\mathbf{y} = \begin{bmatrix} y_1 \\ \vdots \\ y_p \end{bmatrix}$$

The expected value (mean vector) is:

$$E(\mathbf{y}) = \begin{bmatrix} \mu_1 \\ \vdots \\ \mu_p \end{bmatrix}$$

6.1.2 Covariance Matrix

The covariance between any two variables y_i and y_j is:

$$\sigma_{ij} = \text{cov}(y_i, y_j) = E[(y_i - \mu_i)(y_j - \mu_j)]$$

This leads to the **variance-covariance matrix**, also called the **dispersion matrix**:

$$\boldsymbol{\Sigma} = (\sigma_{ij}) = \begin{bmatrix} \sigma_{11} & \sigma_{12} & \ldots & \sigma_{1p} \\ \sigma_{21} & \sigma_{22} & \ldots & \sigma_{2p} \\ \vdots & \vdots & \ddots & \vdots \\ \sigma_{p1} & \sigma_{p2} & \ldots & \sigma_{pp} \end{bmatrix}$$

where $\sigma_{ii} = \text{Var}(y_i)$ represents the variance of y_i. Since covariance is symmetric, we have:

$$\sigma_{ij} = \sigma_{ji}, \quad \forall i, j.$$

If we consider two random vectors $\mathbf{u}_{p\times 1}$ and $\mathbf{v}_{q\times 1}$ with means μ_u and μ_v, their **cross-covariance matrix** is:

$$\boldsymbol{\Sigma}_{uv} = \operatorname{cov}(\mathbf{u}, \mathbf{v}) = E[(\mathbf{u} - \mu_u)(\mathbf{v} - \mu_v)']$$

where $\boldsymbol{\Sigma}_{uv} \neq \boldsymbol{\Sigma}_{vu}$, but they satisfy:

$$\boldsymbol{\Sigma}_{uv} = \boldsymbol{\Sigma}'_{vu}.$$

6.1.2.1 Properties of Covariance Matrices

A valid covariance matrix $\boldsymbol{\Sigma}$ satisfies the following properties:

1. **Symmetry**:

$$\boldsymbol{\Sigma}' = \boldsymbol{\Sigma}.$$

2. **Non-negative definiteness**:

$$\mathbf{a}'\boldsymbol{\Sigma}\mathbf{a} \geq 0, \quad \forall \mathbf{a} \in \mathbb{R}^p,$$

which implies that the eigenvalues $\lambda_1, \ldots, \lambda_p$ satisfy:

$$\lambda_1 \geq \lambda_2 \geq \cdots \geq \lambda_p \geq 0.$$

3. **Generalized variance** (determinant of $\boldsymbol{\Sigma}$):

$$|\boldsymbol{\Sigma}| = \lambda_1\lambda_2 \ldots \lambda_p \geq 0.$$

4. **Total variance** (trace of $\boldsymbol{\Sigma}$):

$$\operatorname{tr}(\boldsymbol{\Sigma}) = \sum_{i=1}^{p} \lambda_i = \sum_{i=1}^{p} \sigma_{ii}.$$

5. **Positive definiteness** (a common assumption in multivariate analysis):

- All eigenvalues of $\boldsymbol{\Sigma}$ are strictly positive.
- $\boldsymbol{\Sigma}$ has an inverse $\boldsymbol{\Sigma}^{-1}$, satisfying:

$$\boldsymbol{\Sigma}^{-1}\boldsymbol{\Sigma} = \mathbf{I}_{p\times p} = \boldsymbol{\Sigma}\boldsymbol{\Sigma}^{-1}.$$

6.1.2.2 Correlation Matrices

The **correlation matrix** provides a standardized measure of linear relationships between variables. The correlation between two variables y_i and y_j is defined as:

$$\rho_{ij} = \frac{\sigma_{ij}}{\sqrt{\sigma_{ii}\sigma_{jj}}}$$

where σ_{ij} is the covariance and σ_{ii} and σ_{jj} are variances.

Thus, the **correlation matrix R** is:

$$\mathbf{R} = \begin{bmatrix} \rho_{11} & \rho_{12} & \cdots & \rho_{1p} \\ \rho_{21} & \rho_{22} & \cdots & \rho_{2p} \\ \vdots & \vdots & \ddots & \vdots \\ \rho_{p1} & \rho_{p2} & \cdots & \rho_{pp} \end{bmatrix}$$

where $\rho_{ii} = 1$ for all i.

Alternatively, the correlation matrix can be expressed as:

$$\mathbf{R} = [\text{diag}(\mathbf{\Sigma})]^{-1/2}\mathbf{\Sigma}[\text{diag}(\mathbf{\Sigma})]^{-1/2}$$

where:

- $\text{diag}(\mathbf{\Sigma})$ is a diagonal matrix with elements σ_{ii} on the diagonal and zeros elsewhere.
- $\mathbf{A}^{1/2}$ (the square root of a symmetric matrix) is a symmetric matrix satisfying $\mathbf{A} = \mathbf{A}^{1/2}\mathbf{A}^{1/2}$.

6.1.3 Equalities in Expectation and Variance

Let:

- $\mathbf{x}$ and $\mathbf{y}$ be random vectors with means μ_x and μ_y and covariance matrices $\mathbf{\Sigma}_x$ and $\mathbf{\Sigma}_y$.
- $\mathbf{A}$ and $\mathbf{B}$ be matrices of constants, and $\mathbf{c}$ and $\mathbf{d}$ be vectors of constants.

Then the following properties hold:

1. **Expectation transformations**:

$$E(\mathbf{Ay} + \mathbf{c}) = \mathbf{A}\mu_y + \mathbf{c}$$

2. **Variance transformations**:

$$\text{Var}(\mathbf{Ay} + \mathbf{c}) = \mathbf{A}\text{Var}(\mathbf{y})\mathbf{A}' = \mathbf{A}\boldsymbol{\Sigma}_{\mathbf{y}}\mathbf{A}'$$

3. **Covariance of linear transformations**:

$$\text{Cov}(\mathbf{Ay} + \mathbf{c}, \mathbf{By} + \mathbf{d}) = \mathbf{A}\boldsymbol{\Sigma}_{\mathbf{y}}\mathbf{B}'$$

4. **Expectation of combined variables**:

$$E(\mathbf{Ay} + \mathbf{Bx} + \mathbf{c}) = \mathbf{A}\mu_y + \mathbf{B}\mu_x + \mathbf{c}$$

5. **Variance of combined variables**:

$$\text{Var}(\mathbf{Ay} + \mathbf{Bx} + \mathbf{c}) = \mathbf{A}\boldsymbol{\Sigma}_{\mathbf{y}}\mathbf{A}' + \mathbf{B}\boldsymbol{\Sigma}_{\mathbf{x}}\mathbf{B}' + \mathbf{A}\boldsymbol{\Sigma}_{\mathbf{yx}}\mathbf{B}' + \mathbf{B}\boldsymbol{\Sigma}_{\mathbf{yx}}'\mathbf{A}'$$

6.1.4 Multivariate Normal Distribution

The **multivariate normal distribution (MVN)** is fundamental in multivariate analysis. Let $\mathbf{y}$ be a multivariate normal random variable with mean μ and covariance matrix $\boldsymbol{\Sigma}$. Then its **probability density function (PDF)** is:

$$f(\mathbf{y}) = \frac{1}{(2\pi)^{p/2}|\boldsymbol{\Sigma}|^{1/2}} \exp\left(-\frac{1}{2}(\mathbf{y} - \mu)'\boldsymbol{\Sigma}^{-1}(\mathbf{y} - \mu)\right).$$

We denote this distribution as:

$$\mathbf{y} \sim N_p(\mu, \boldsymbol{\Sigma}).$$

6.1.4.1 Properties of the Multivariate Normal Distribution

The multivariate normal distribution has several important properties that are fundamental to multivariate statistical methods.

- **Linear Transformations**:
  Let $\mathbf{A}_{r \times p}$ be a fixed matrix. Then:

$$\mathbf{Ay} \sim N_r(\mathbf{A}\mu, \mathbf{A}\boldsymbol{\Sigma}\mathbf{A}')$$

  where $r \leq p$. Additionally, for $\mathbf{A}\boldsymbol{\Sigma}\mathbf{A}'$ to be non-singular, the rows of $\mathbf{A}$ must be linearly independent.

- **Standardization Using Precision Matrix**:
  Let **G** be a matrix such that:

$$\mathbf{\Sigma}^{-1} = \mathbf{G}\mathbf{G}'$$

  Then:

$$\mathbf{G}'\mathbf{y} \sim N_p(\mathbf{G}'\boldsymbol{\mu}, \mathbf{I})$$

  and:

$$\mathbf{G}'(\mathbf{y} - \boldsymbol{\mu}) \sim N_p(0, \mathbf{I}).$$

  This transformation whitens the data, converting it into an identity covariance structure.
- **Linear Combinations**:
  Any fixed linear combination of $y_1, \ldots, y_p$, say $\mathbf{c}'\mathbf{y}$, follows:

$$\mathbf{c}'\mathbf{y} \sim N_1(\mathbf{c}'\boldsymbol{\mu}, \mathbf{c}'\mathbf{\Sigma}\mathbf{c}).$$

6.1.4.2 Partitioning the MVN Distribution

Consider a partitioned random vector:

$$\mathbf{y} = \begin{bmatrix} \mathbf{y}_1 \\ \mathbf{y}_2 \end{bmatrix} \sim N_p \left(\begin{bmatrix} \mu_1 \\ \mu_2 \end{bmatrix}, \begin{bmatrix} \mathbf{\Sigma}_{11} & \mathbf{\Sigma}_{12} \\ \mathbf{\Sigma}_{21} & \mathbf{\Sigma}_{22} \end{bmatrix} \right).$$

where:

- $\mathbf{y}_1$ is $p_1 \times 1$.
- $\mathbf{y}_2$ is $p_2 \times 1$.
- $p_1 + p_2 = p$.
- $p_1, p_2 \geq 1$.

  The marginal distributions of $\mathbf{y}_1$ and $\mathbf{y}_2$ are:

$$\mathbf{y}_1 \sim N_{p_1}(\boldsymbol{\mu}_1, \mathbf{\Sigma}_{11}) \quad \text{and} \quad \mathbf{y}_2 \sim N_{p_2}(\boldsymbol{\mu}_2, \mathbf{\Sigma}_{22}).$$

  Each component y_i follows:

$$y_i \sim N_1(\mu_i, \sigma_{ii}).$$

  The conditional distribution of $\mathbf{y}_1$ given $\mathbf{y}_2$ is also normal:

$$\mathbf{y}_1|\mathbf{y}_2 \sim N_{p_1}\left(\boldsymbol{\mu}_1 + \mathbf{\Sigma}_{12}\mathbf{\Sigma}_{22}^{-1}(\mathbf{y}_2 - \boldsymbol{\mu}_2), \mathbf{\Sigma}_{11} - \mathbf{\Sigma}_{12}\mathbf{\Sigma}_{22}^{-1}\mathbf{\Sigma}_{21}\right).$$

This equation shows that knowing $\mathbf{y}_2$ adjusts the mean of $\mathbf{y}_1$, and the variance is reduced.

Similarly, the conditional distribution of $\mathbf{y}_2$ given $\mathbf{y}_1$ follows the same structure.

- $\mathbf{y}_1$ and $\mathbf{y}_2$ are independent if and only if:

$$\boldsymbol{\Sigma}_{12} = 0.$$

If $\mathbf{y} \sim N(\boldsymbol{\mu}, \boldsymbol{\Sigma})$ and $\boldsymbol{\Sigma}$ is positive definite, then:

$$(\mathbf{y} - \mu)'\boldsymbol{\Sigma}^{-1}(\mathbf{y} - \mu) \sim \chi^2_p.$$

This property is essential in hypothesis testing and Mahalanobis distance calculations.

6.1.4.3 Summation of Independent MVN Variables

If $\mathbf{y}_i$ are independent random vectors following:

$$\mathbf{y}_i \sim N_p(\boldsymbol{\mu}_i, \boldsymbol{\Sigma}_i),$$

then for fixed matrices $\mathbf{A}_{i(m\times p)}$, the sum:

$$\sum_{i=1}^{k} \mathbf{A}_i \mathbf{y}_i$$

follows:

$$\sum_{i=1}^{k} \mathbf{A}_i \mathbf{y}_i \sim N_m\Big(\sum_{i=1}^{k} \mathbf{A}_i \boldsymbol{\mu}_i, \sum_{i=1}^{k} \mathbf{A}_i \boldsymbol{\Sigma}_i \mathbf{A}_i'\Big).$$

This property underpins multivariate regression and linear discriminant analysis.

6.1.4.4 Multiple Regression

In multivariate analysis, multiple regression extends simple regression to cases where multiple predictor variables influence a response variable. Suppose:

$$\begin{pmatrix} Y \\ \mathbf{x} \end{pmatrix} \sim N_{p+1}\left(\begin{bmatrix} \mu_y \\ \boldsymbol{\mu}_x \end{bmatrix}, \begin{bmatrix} \sigma_Y^2 & \boldsymbol{\Sigma}_{yx} \\ \boldsymbol{\Sigma}_{yx} & \boldsymbol{\Sigma}_{xx} \end{bmatrix}\right)$$

where:

- Y is a scalar response variable.
- $\mathbf{x}$ is a $p \times 1$ vector of predictors.
- μ_y and $\boldsymbol{\mu}_x$ are the respective means.
- σ_Y^2 is the variance of Y.
- $\boldsymbol{\Sigma}_{xx}$ is the covariance matrix of $\mathbf{x}$.
- $\boldsymbol{\Sigma}_{yx}$ is the covariance vector between Y and $\mathbf{x}$.

From the properties of the multivariate normal distribution, the conditional expectation of Y given $\mathbf{x}$ is:

$$\begin{aligned} E(Y|\mathbf{x}) &= \mu_y + \boldsymbol{\Sigma}_{yx}\boldsymbol{\Sigma}_{xx}^{-1}(\mathbf{x} - \boldsymbol{\mu}_x) \\ &= \mu_y - \boldsymbol{\Sigma}_{yx}\boldsymbol{\Sigma}_{xx}^{-1}\boldsymbol{\mu}_x + \boldsymbol{\Sigma}_{yx}\boldsymbol{\Sigma}_{xx}^{-1}\mathbf{x} \\ &= \beta_0 + \boldsymbol{\beta}'\mathbf{x}, \end{aligned}$$

where:

- $\beta_0 = \mu_y - \boldsymbol{\Sigma}_{yx}\boldsymbol{\Sigma}_{xx}^{-1}\boldsymbol{\mu}_x$ (intercept).
- $\boldsymbol{\beta} = (\beta_1, \ldots, \beta_p)' = \boldsymbol{\Sigma}_{xx}^{-1}\boldsymbol{\Sigma}_{yx}'$ (regression coefficients).

This resembles the least squares estimator:

$$\boldsymbol{\beta} = (\mathbf{x}'\mathbf{x})^{-1}\mathbf{x}'\mathbf{y},$$

but differs when considering the theoretical covariance relationships rather than empirical estimates.

The conditional variance of Y given $\mathbf{x}$ is:

$$\text{Var}(Y|\mathbf{x}) = \sigma_Y^2 - \boldsymbol{\Sigma}_{yx}\boldsymbol{\Sigma}_{xx}^{-1}\boldsymbol{\Sigma}_{yx}'.$$

This shows that knowing $\mathbf{x}$ reduces uncertainty in predicting Y.

6.1.4.5 Samples from Multivariate Normal Populations

Suppose we have a random sample of size n, denoted as:

$$\mathbf{y}_1, \ldots, \mathbf{y}_n \sim N_p(\boldsymbol{\mu}, \boldsymbol{\Sigma}).$$

Then:

1. **Sample Mean**: The sample mean is given by:

$$\bar{\mathbf{y}} = \frac{1}{n}\sum_{i=1}^{n}\mathbf{y}_i.$$

Since $\mathbf{y}_i$ are independent and identically distributed (iid), it follows that:

$$\bar{\mathbf{y}} \sim N_p(\boldsymbol{\mu}, \boldsymbol{\Sigma}/n).$$

This implies that $\bar{\mathbf{y}}$ is an unbiased estimator of $\boldsymbol{\mu}$.

2. **Sample Covariance Matrix**: The $p \times p$ sample variance-covariance matrix is:

$$\mathbf{S} = \frac{1}{n-1} \sum_{i=1}^{n} (\mathbf{y}_i - \bar{\mathbf{y}})(\mathbf{y}_i - \bar{\mathbf{y}})'.$$

Expanding this:

$$\mathbf{S} = \frac{1}{n-1} \left(\sum_{i=1}^{n} \mathbf{y}_i \mathbf{y}_i' - n\bar{\mathbf{y}}\bar{\mathbf{y}}' \right).$$

- $\mathbf{S}$ is symmetric.
- $\mathbf{S}$ is an unbiased estimator of $\boldsymbol{\Sigma}$.
- $\mathbf{S}$ contains $p(p+1)/2$ unique random variables.

3. **Wishart Distribution**: The scaled sample covariance matrix follows a Wishart distribution:

$$(n-1)\mathbf{S} \sim W_p(n-1, \boldsymbol{\Sigma}).$$

where:

- $W_p(n-1, \boldsymbol{\Sigma})$ is a Wishart distribution with $n-1$ degrees of freedom.
- $E[(n-1)\mathbf{S}] = (n-1)\boldsymbol{\Sigma}$.

The Wishart distribution is a multivariate generalization of the chi-square distribution.

4. **Independence of $\bar{\mathbf{y}}$ and S**: The sample mean $\bar{\mathbf{y}}$ and sample covariance matrix $\mathbf{S}$ are independent:

$$\bar{\mathbf{y}} \perp \mathbf{S}.$$

This result is crucial for inference in multivariate hypothesis testing.

5. **Sufficiency of $\bar{\mathbf{y}}$ and S**: The pair $(\bar{\mathbf{y}}, \mathbf{S})$ are sufficient statistics for $(\boldsymbol{\mu}, \boldsymbol{\Sigma})$. That is, all the information about $\boldsymbol{\mu}$ and $\boldsymbol{\Sigma}$ in the sample is contained in $\bar{\mathbf{y}}$ and $\mathbf{S}$, regardless of sample size.

6.1.4.6 Large Sample Properties

Consider a random sample $\mathbf{y}_1, \ldots, \mathbf{y}_n$ drawn from a population with mean $\boldsymbol{\mu}$ and variance-covariance matrix $\boldsymbol{\Sigma}$.

Key Properties

- **Consistency of Estimators**:
  - The sample mean $\bar{\mathbf{y}}$ is a consistent estimator of $\boldsymbol{\mu}$.
  - The sample covariance matrix $\mathbf{S}$ is a consistent estimator of $\boldsymbol{\Sigma}$.
- Multivariate [Central Limit Theorem]:
  - Similar to the univariate case, the sample mean follows approximately:

$$\sqrt{n}(\bar{\mathbf{y}} - \mu)\dot{\sim}N_p(\mathbf{0}, \boldsymbol{\Sigma})$$

    This approximation holds when the sample size is large relative to the number of variables ($n \geq 25p$).
  - Equivalently, the sample mean follows:

$$\bar{\mathbf{y}}\dot{\sim}N_p(\boldsymbol{\mu}, \boldsymbol{\Sigma}/n).$$

- **Wald's Theorem**:
  - When n is large relative to p:

$$n(\bar{\mathbf{y}} - \boldsymbol{\mu})'\mathbf{S}^{-1}(\bar{\mathbf{y}} - \boldsymbol{\mu}) \sim \chi^2_p.$$

    This is useful for hypothesis testing about $\boldsymbol{\mu}$.

6.1.4.7 Maximum Likelihood Estimation for MVN

Suppose $\mathbf{y}_1, \ldots, \mathbf{y}_n$ are iid random vectors from:

$$\mathbf{y}_i \sim N_p(\boldsymbol{\mu}, \boldsymbol{\Sigma}).$$

The likelihood function for the sample is:

$$\begin{aligned} L(\boldsymbol{\mu}, \boldsymbol{\Sigma}) &= \prod_{j=1}^{n} \left[\frac{1}{(2\pi)^{p/2}|\boldsymbol{\Sigma}|^{1/2}} \exp\left(-\frac{1}{2}(\mathbf{y}_j - \boldsymbol{\mu})'\boldsymbol{\Sigma}^{-1}(\mathbf{y}_j - \boldsymbol{\mu}) \right) \right] \\ &= \frac{1}{(2\pi)^{np/2}|\boldsymbol{\Sigma}|^{n/2}} \exp\left(-\frac{1}{2}\sum_{j=1}^{n}(\mathbf{y}_j - \boldsymbol{\mu})'\boldsymbol{\Sigma}^{-1}(\mathbf{y}_j - \boldsymbol{\mu}) \right). \end{aligned}$$

Taking the log-likelihood function and differentiating with respect to $\boldsymbol{\mu}$ and $\boldsymbol{\Sigma}$ leads to the maximum likelihood estimators:

The MLE for the mean is simply the sample mean:

$$\hat{\boldsymbol{\mu}} = \bar{\mathbf{y}}.$$

The MLE for the covariance matrix is:

$$\hat{\boldsymbol{\Sigma}} = \frac{n-1}{n}\mathbf{S}.$$

where:

$$\mathbf{S} = \frac{1}{n-1}\sum_{j=1}^{n}(\mathbf{y}_j - \bar{\mathbf{y}})(\mathbf{y}_j - \bar{\mathbf{y}})'.$$

This differs from $\mathbf{S}$ by the factor $\frac{n-1}{n}$, making $\hat{\boldsymbol{\Sigma}}$ a **biased estimator** of $\boldsymbol{\Sigma}$.

6.1.4.7.1 Properties of Maximum Likelihood Estimators

MLEs have several important theoretical properties:

1. **Invariance**:
   - If $\hat{\theta}$ is the MLE of θ, then the MLE of any function $h(\theta)$ is:

   $$h(\hat{\theta}).$$

2. **Consistency**:
   - MLEs are consistent estimators, meaning they converge to the true parameter values as $n \to \infty$.
   - However, they can be biased for finite samples.
3. **Efficiency**:
   - MLEs are asymptotically efficient, meaning they achieve the Cramér-Rao lower bound for variance in large samples.
   - No other estimator has a smaller variance asymptotically.
4. **Asymptotic Normality**:
   - Suppose $\hat{\theta}_n$ is the MLE for θ based on n independent observations.
   - Then, for large n:

   $$\hat{\theta}_n \dot{\sim} N(\theta, \mathbf{H}^{-1}),$$

where **H** is the [Fisher Information Matrix], defined as:

$$\mathbf{H}_{ij} = -E\left(\frac{\partial^2 l(\boldsymbol{\theta})}{\partial\theta_i \partial\theta_j}\right).$$

- The [Fisher Information Matrix] measures the amount of information in the data about θ.
- It can be estimated by evaluating the second derivatives of the log-likelihood function at $\hat{\theta}_n$.

6.1.4.7.2 Likelihood Ratio Testing

MLEs allow us to construct likelihood ratio tests for hypothesis testing.

- Suppose we test a null hypothesis H_0:

$$H_0 : \boldsymbol{\theta} \in \Theta_0 \quad \text{vs.} \quad H_A : \boldsymbol{\theta} \in \Theta.$$

- The likelihood ratio statistic is:

$$\Lambda = \frac{\max_{\theta \in \Theta_0} L(\boldsymbol{\mu}, \boldsymbol{\Sigma}|\mathbf{Y})}{\max_{\theta \in \Theta} L(\boldsymbol{\mu}, \boldsymbol{\Sigma}|\mathbf{Y})}.$$

- Under large sample conditions, we use the Wilks' theorem, which states:

$$-2\log\Lambda \sim \chi^2_v,$$

  where:

  - v is the difference in the number of parameters between the unrestricted and restricted models.
  - This allows us to approximate the distribution of $-2\log\Lambda$ using the chi-square distribution.

6.1.5 *Test of Multivariate Normality*

Assessing multivariate normality is essential for many statistical techniques, including multivariate regression, principal component analysis, and MANOVA. Below are key methods for testing MVN.

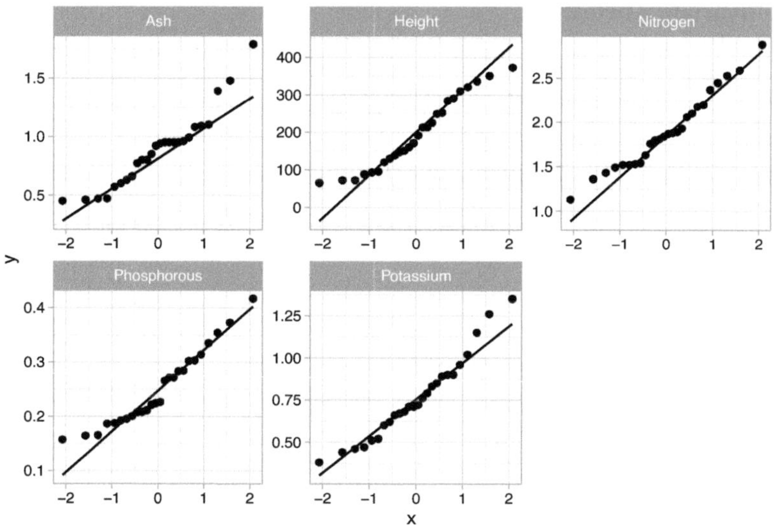

Fig. 6.1 Scatterplot

6.1.5.1 Univariate Normality Checks

Before testing for multivariate normality, it is useful to check for univariate normality in each variable separately (Fig. 6.1):

- Normality Assessment: Visual and statistical tests can be used to check normality.
- Key Property: If any univariate distribution is not normal, then the joint multivariate distribution cannot be normal.
- Important Caveat: Even if all univariate distributions are normal, this does not guarantee multivariate normality.

Thus, **univariate normality is a necessary but not sufficient condition** for MVN.

6.1.5.2 Mardia's Test for Multivariate Normality

Mardia (1970) proposed two measures for assessing MVN:

1. **Multivariate Skewness**
   Defined as:

$$\beta_{1,p} = E[(\mathbf{y} - \boldsymbol{\mu})'\boldsymbol{\Sigma}^{-1}(\mathbf{x} - \boldsymbol{\mu})]^3,$$

   where **x** and **y** are independent but identically distributed.

2. Multivariate Kurtosis
Defined as:

$$\beta_{2,p} = E[(\mathbf{y} - \boldsymbol{\mu})'\boldsymbol{\Sigma}^{-1}(\mathbf{x} - \boldsymbol{\mu})]^2.$$

For a **true multivariate normal distribution**:

$$\beta_{1,p} = 0, \quad \beta_{2,p} = p(p+2).$$

Sample Estimates
For a random sample of size n, we estimate:

$$\hat{\beta}_{1,p} = \frac{1}{n^2} \sum_{i=1}^{n} \sum_{j=1}^{n} g_{ij}^2,$$

$$\hat{\beta}_{2,p} = \frac{1}{n} \sum_{i=1}^{n} g_{ii}^2,$$

where:

- $g_{ij} = (\mathbf{y}_i - \bar{\mathbf{y}})'\mathbf{S}^{-1}(\mathbf{y}_j - \bar{\mathbf{y}})$,
- $g_{ii} = d_i^2$, which is the Mahalanobis distance.

Mardia (1970) derived the following large-sample approximations:

$$\kappa_1 = \frac{n\hat{\beta}_{1,p}}{6} \dot{\sim} \chi^2_{p(p+1)(p+2)/6},$$

$$\kappa_2 = \frac{\hat{\beta}_{2,p} - p(p+2)}{\sqrt{8p(p+2)/n}} \sim N(0, 1).$$

Interpretation

- κ_1 and κ_2 are test statistics for the null hypothesis of MVN.
- Non-normality in means is associated with skewness ($\beta_{1,p}$).
- Non-normality in covariance is associated with kurtosis ($\beta_{2,p}$).

See Fig. 6.2 for an example of the Mardia's test for multivariate normality.

6.1.5.3 Doornik-Hansen Test

- This test transforms variables to approximate normality using skewness and kurtosis corrections (Doornik and Hansen, 2008).
- Recommended when sample sizes are small.

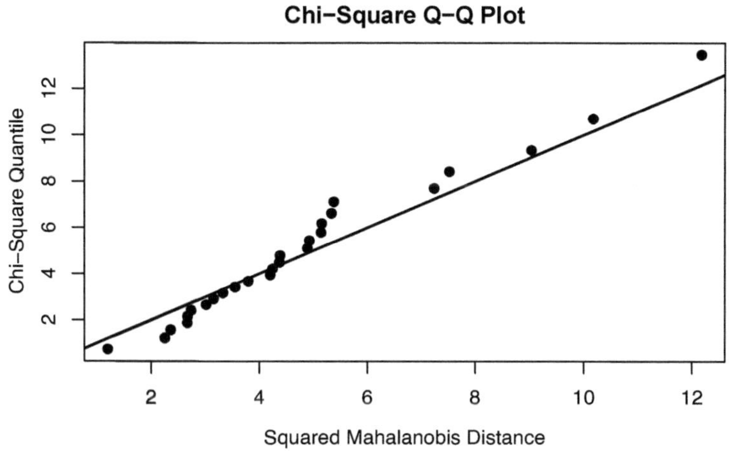

Fig. 6.2 Chi-square Q-Q plot Mardia

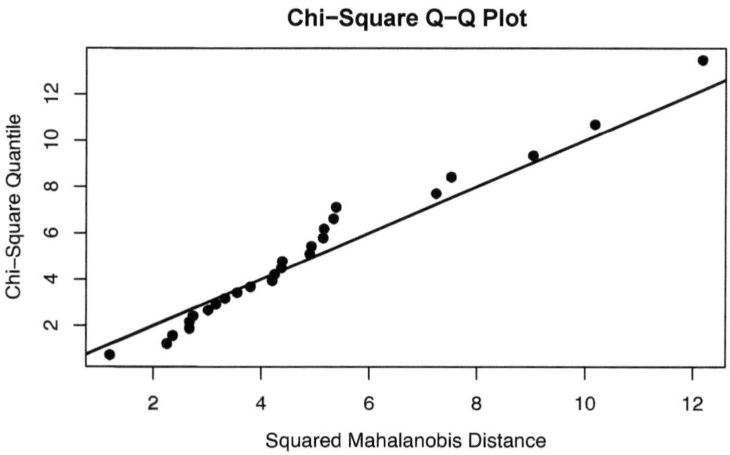

Fig. 6.3 Chi-square Q-Q plot Doornik-Hansen test

See Fig. 6.3 for an example of the Doornik-Hansen test.

6.1.5.4 Chi-Square Q-Q Plot

The Chi-Square Q-Q plot is a graphical method for assessing MVN:

1. Compute Mahalanobis distances:

$$d_i^2 = (\mathbf{y}_i - \bar{\mathbf{y}})'\mathbf{S}^{-1}(\mathbf{y}_i - \bar{\mathbf{y}}).$$

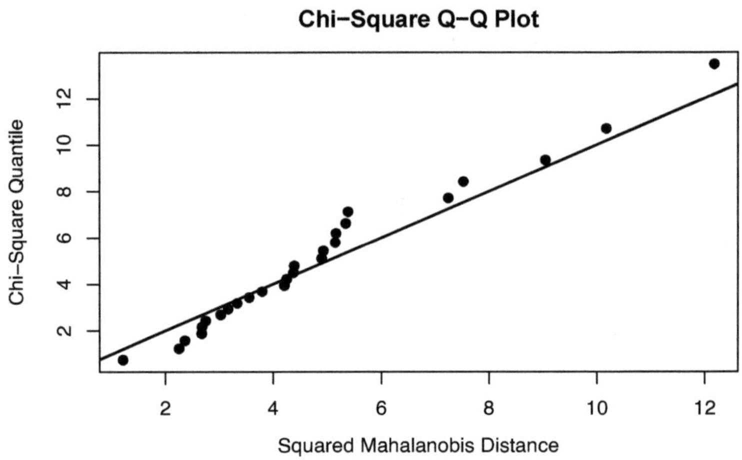

Fig. 6.4 Chi-square Q-Q plot Henze-Zirkler test

2. The transformed variables:

$$\mathbf{z}_i = \boldsymbol{\Sigma}^{-1/2}(\mathbf{y}_i - \boldsymbol{\mu})$$

are iid from $N_p(\mathbf{0}, \mathbf{I})$, and thus:

$$d_i^2 \sim \chi_p^2.$$

3. Plot ordered d_i^2 values against the theoretical quantiles of the χ_p^2 distribution.

Interpretation

- If the data are MVN, the plot should resemble a straight line at 45°.
- Deviations suggest non-normality, especially in the tails.

Limitations

- Requires a large sample size.
- Even when data are truly MVN, the tails may deviate.

For other tests, such as the Henze-Zirkler test, Royston's test, and Energy test, see Figs. 6.4, 6.5, and 6.6.

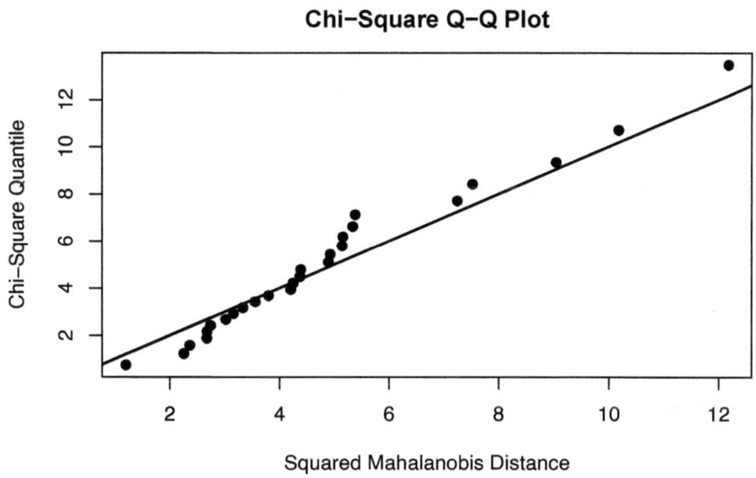

Fig. 6.5 Chi-square Q-Q Plot Royston test

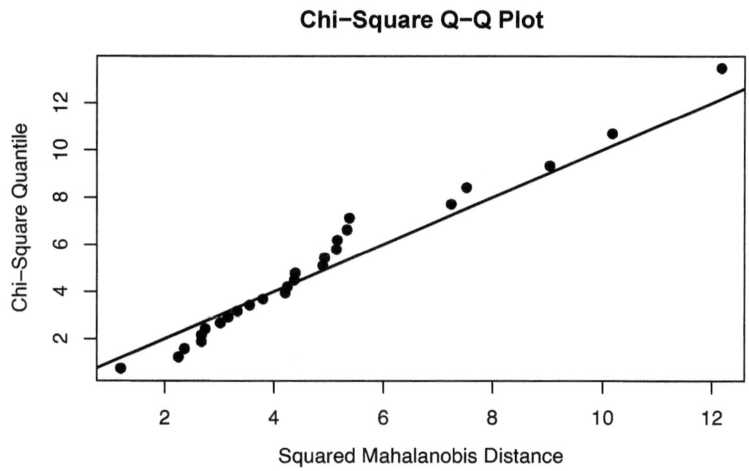

Fig. 6.6 Chi-square Q-Q plot energy test

6.1.5.5 Handling Non-normality

If data **fail** the multivariate normality tests, possible approaches include:

1. **Ignoring non-normality** (acceptable for large samples due to the [Central Limit Theorem])
2. **Using nonparametric methods** (e.g., permutation tests)
3. **Applying approximate models** (e.g., [Generalized Linear Mixed Models])
4. **Transforming the data** (e.g., log, Box-Cox, or rank transformations 6.3)

```
# Load necessary libraries
library(heplots)      # Multivariate hypothesis tests
library(ICSNP)        # Multivariate tests
library(MVN)          # Multivariate normality tests
library(tidyverse)    # Data wrangling & visualization


# Load dataset
trees <- read.table("images/trees.dat")
names(trees) <-
    c("Nitrogen", "Phosphorous", "Potassium", "Ash", "Height")

# Structure of dataset
str(trees)
#> 'data.frame':    26 obs. of  5 variables:
#>  $ Nitrogen   : num  2.2 2.1 1.52 2.88 2.18 1.87 1.52 2.37
↪  2.06 1.84 ...
#>  $ Phosphorous: num  0.417 0.354 0.208 0.335 0.314 0.271
↪  0.164 0.302 0.373 0.265 ...
#>  $ Potassium  : num  1.35 0.9 0.71 0.9 1.26 1.15 0.83 0.89
↪  0.79 0.72 ...
#>  $ Ash        : num  1.79 1.08 0.47 1.48 1.09 0.99 0.85
↪  0.94 0.8 0.77 ...
#>  $ Height     : int  351 249 171 373 321 191 225 291 284
↪  213 ...

# Summary statistics
summary(trees)
#>     Nitrogen       Phosphorous        Potassium           Ash
↪
#>  Min.   :1.130   Min.   :0.1570   Min.   :0.3800   Min.
↪  :0.4500
#>  1st Qu.:1.532   1st Qu.:0.1963   1st Qu.:0.6050   1st
↪  Qu.:0.6375
```

```
#>  Median :1.855   Median :0.2250   Median :0.7150   Median
↪   :0.9300
#>  Mean   :1.896   Mean   :0.2506   Mean   :0.7619   Mean
↪   :0.8873
#>  3rd Qu.:2.160   3rd Qu.:0.2975   3rd Qu.:0.8975   3rd
↪   Qu.:0.9825
#>  Max.   :2.880   Max.   :0.4170   Max.   :1.3500   Max.
↪   :1.7900
#>      Height
#>  Min.   : 65.0
#>  1st Qu.:122.5
#>  Median :181.0
#>  Mean   :196.6
#>  3rd Qu.:276.0
#>  Max.   :373.0

# Pearson correlation matrix
cor(trees, method = "pearson")
#>              Nitrogen Phosphorous Potassium       Ash
↪   Height
#> Nitrogen    1.0000000   0.6023902 0.5462456 0.6509771
↪   0.8181641
#> Phosphorous 0.6023902   1.0000000 0.7037469 0.6707871
↪   0.7739656
#> Potassium   0.5462456   0.7037469 1.0000000 0.6710548
↪   0.7915683
#> Ash         0.6509771   0.6707871 0.6710548 1.0000000
↪   0.7676771
#> Height      0.8181641   0.7739656 0.7915683 0.7676771
↪   1.0000000

# Q-Q plots for each variable
gg <- trees %>%
    pivot_longer(everything(), names_to = "Var", values_to =
    ↪  "Value") %>%
    ggplot(aes(sample = Value)) +
    geom_qq() +
    geom_qq_line() +
    facet_wrap( ~ Var, scales = "free")
```

```r
print(gg)
```

```r
# Shapiro-Wilk test for univariate normality
sw_tests <- apply(trees, MARGIN = 2, FUN = shapiro.test)
sw_tests
#> $Nitrogen
#>
#>  Shapiro-Wilk normality test
#>
#> data:  newX[, i]
#> W = 0.96829, p-value = 0.5794
#>
#>
#> $Phosphorous
#>
#>  Shapiro-Wilk normality test
#>
#> data:  newX[, i]
#> W = 0.93644, p-value = 0.1104
#>
#>
#> $Potassium
#>
#>  Shapiro-Wilk normality test
#>
#> data:  newX[, i]
#> W = 0.95709, p-value = 0.3375
#>
#>
#> $Ash
#>
#>  Shapiro-Wilk normality test
#>
#> data:  newX[, i]
#> W = 0.92071, p-value = 0.04671
#>
#>
#> $Height
#>
#>  Shapiro-Wilk normality test
#>
#> data:  newX[, i]
#> W = 0.94107, p-value = 0.1424
```

```
# Kolmogorov-Smirnov test for normality
ks_tests <- map(trees, ~ ks.test(scale(.x), "pnorm"))
ks_tests
#>  $Nitrogen
#>
#>  Asymptotic one-sample Kolmogorov-Smirnov test
#>
#> data:  scale(.x)
#> D = 0.12182, p-value = 0.8351
#> alternative hypothesis: two-sided
#>
#>
#> $Phosphorous
#>
#>  Asymptotic one-sample Kolmogorov-Smirnov test
#>
#> data:  scale(.x)
#> D = 0.17627, p-value = 0.3944
#> alternative hypothesis: two-sided
#>
#>
#> $Potassium
#>
#>  Asymptotic one-sample Kolmogorov-Smirnov test
#>
#> data:  scale(.x)
#> D = 0.10542, p-value = 0.9348
#> alternative hypothesis: two-sided
#>
#>
#> $Ash
#>
#>  Asymptotic one-sample Kolmogorov-Smirnov test
#>
#> data:  scale(.x)
#> D = 0.14503, p-value = 0.6449
#> alternative hypothesis: two-sided
#>
#>
#> $Height
#>
#>  Asymptotic one-sample Kolmogorov-Smirnov test
#>
```

```
#> data:  scale(.x)
#> D = 0.1107, p-value = 0.9076
#> alternative hypothesis: two-sided
```

```
# Mardia's test for multivariate normality
mardia_test <-
    mvn(
        trees,
        mvnTest = "mardia",
        covariance = FALSE,
        multivariatePlot = "qq"
    )
```

```
mardia_test$multivariateNormality
#>              Test          Statistic           p value
↪  Result
#> 1 Mardia Skewness  29.7248528871795   0.72054426745778
↪  YES
#> 2 Mardia Kurtosis -1.67743173185383 0.0934580886477281
↪  YES
#> 3             MVN              <NA>               <NA>
↪  YES
```

```
# Doornik-Hansen test
dh_test <-
    mvn(
        trees,
        mvnTest = "dh",
        covariance = FALSE,
        multivariatePlot = "qq"
    )
```

```
dh_test$multivariateNormality
#>              Test        E df      p value MVN
#> 1 Doornik-Hansen 161.9446 10 1.285352e-29  NO
```

```
# Henze-Zirkler test
hz_test <-
    mvn(
        trees,
        mvnTest = "hz",
        covariance = FALSE,
        multivariatePlot = "qq"
    )
```

```
hz_test$multivariateNormality
#>            Test        HZ   p value MVN
#> 1 Henze-Zirkler 0.7591525 0.6398905 YES
```

```
# Royston's test (only for 3 < obs < 5000)
royston_test <-
    mvn(
        trees,
        mvnTest = "royston",
        covariance = FALSE,
        multivariatePlot = "qq"
    )
```

```
royston_test$multivariateNormality
#>      Test        H    p value MVN
#> 1 Royston 9.064631 0.08199215 YES
```

```
# Energy test
estat_test <-
    mvn(
        trees,
        mvnTest = "energy",
        covariance = FALSE,
        multivariatePlot = "qq"
    )
```

```
estat_test$multivariateNormality
#>          Test Statistic p value MVN
#> 1 E-statistic  1.091101   0.528 YES
```

6.1.6 Mean Vector Inference

6.1.6.1 Univariate Case

In the univariate normal distribution, we test:

$$H_0 : \mu = \mu_0$$

using the t-test statistic:

$$T = \frac{\bar{y} - \mu_0}{s/\sqrt{n}} \sim t_{n-1}.$$

Decision Rule

- If H_0 is true, then T follows a t-distribution with $n - 1$ degrees of freedom.
- We reject H_0 if:

$$|T| > t_{(1-\alpha/2,n-1)}$$

  because an extreme value suggests that observing $\bar{y}$ under H_0 is unlikely.

Alternative Formulation

Squaring T, we obtain:

$$T^2 = \frac{(\bar{y} - \mu_0)^2}{s^2/n} = n(\bar{y} - \mu_0)(s^2)^{-1}(\bar{y} - \mu_0).$$

Under H_0:

$$T^2 \sim f_{(1,n-1)}.$$

This formulation allows for a direct extension to the multivariate case.

6.1.6.2 Multivariate Generalization: Hotelling's T^2 Test

For a p-dimensional mean vector, we test:

$$H_0 : \boldsymbol{\mu} = \boldsymbol{\mu}_0,$$
$$H_a : \boldsymbol{\mu} \neq \boldsymbol{\mu}_0.$$

Define the Hotelling's T^2 test statistic:

$$T^2 = n(\bar{\mathbf{y}} - \boldsymbol{\mu}_0)'\mathbf{S}^{-1}(\bar{\mathbf{y}} - \boldsymbol{\mu}_0).$$

where:

- $\bar{\mathbf{y}}$ is the sample mean vector.
- $\mathbf{S}$ is the sample covariance matrix.
- T^2 can be interpreted as a generalized squared distance between $\bar{\mathbf{y}}$ and $\boldsymbol{\mu}_0$.

Under multivariate normality, the test statistic follows an F-distribution:

$$F = \frac{n-p}{(n-1)p}T^2 \sim f_{(p,n-p)}.$$

We reject H_0 if:

$$F > f_{(1-\alpha,p,n-p)}.$$

Key Properties of Hotelling's T^2 Test

1. **Invariance to Measurement Scale**:

   - If we apply a linear transformation to the data:

   $$\mathbf{z} = \mathbf{C}\mathbf{y} + \mathbf{d},$$

   where $\mathbf{C}$ and $\mathbf{d}$ do not depend on $\mathbf{y}$, then:

   $$T^2(\mathbf{z}) = T^2(\mathbf{y}).$$

   This ensures that unit changes (e.g., inches to centimeters) do not affect the test results.

2. **Likelihood Ratio Test**:

   - The T^2 test can be derived as a likelihood ratio test for $H_0 : \boldsymbol{\mu} = \boldsymbol{\mu}_0$.

```r
# Load required packages
library(MASS)     # For multivariate analysis
library(ICSNP)    # For Hotelling's T^2 test

# Simulated dataset (5 variables, 30 observations)
set.seed(123)
n <- 30  # Sample size
p <- 5   # Number of variables
mu <- rep(0, p)  # Population mean vector
Sigma <- diag(p) # Identity covariance matrix

# Generate multivariate normal data
data <- mvrnorm(n, mu, Sigma)
colnames(data) <- paste0("V", 1:p)

# Compute sample mean and covariance
sample_mean <- colMeans(data)
sample_cov  <- cov(data)

# Perform Hotelling's T^2 test (testing against mu_0 = rep(0,
↪  p))
hotelling_test <- HotellingsT2(data, mu = rep(0, p))

# Print results
print(hotelling_test)
#>
#>  Hotelling's one sample T2-test
#>
#> data:  data
#> T.2 = 0.43475, df1 = 5, df2 = 25, p-value = 0.82
#> alternative hypothesis: true location is not equal to
↪  c(0,0,0,0,0)
```

6.1.6.3 Confidence Intervals

6.1.6.3.1 Confidence Region for the Mean Vector

An exact $100(1-\alpha)\%$ confidence region for the population mean vector $\boldsymbol{\mu}$ is the set of all vectors $\mathbf{v}$ that are "close enough" to the observed mean vector $\bar{\mathbf{y}}$ such that:

$$n(\bar{\mathbf{y}} - \boldsymbol{\mu}_0)'\mathbf{S}^{-1}(\bar{\mathbf{y}} - \boldsymbol{\mu}_0) \le \frac{(n-1)p}{n-p} f_{(1-\alpha,p,n-p)}.$$

Interpretation

- The confidence region consists of all mean vectors $\boldsymbol{\mu}_0$ for which we fail to reject H_0 in the Hotelling's T^2 test.
- If $p = 2$, this confidence region forms a hyper-ellipsoid.

Why Use Confidence Regions?

- They provide a joint assessment of plausible values for $\boldsymbol{\mu}$.
- However, in practice, we often prefer individual confidence intervals for each mean component.

6.1.6.3.2 Simultaneous Confidence Intervals

We want simultaneous confidence statements, ensuring that all individual confidence intervals hold simultaneously with high probability.

Simultaneous Confidence Intervals (General Form)
By projecting the confidence region onto the coordinate axes, we obtain simultaneous confidence intervals:

$$\bar{y}_i \pm \sqrt{\frac{(n-1)p}{n-p} f_{(1-\alpha, p, n-p)} \frac{s_{ii}}{n}}, \quad \text{for } i = 1, \dots, p.$$

- These intervals are conservative, meaning their actual confidence level is at least $100(1 - \alpha)\%$.

Simultaneous Confidence Intervals for Any Linear Combination
For any arbitrary linear combination $\mathbf{a}'\boldsymbol{\mu}$:

$$\mathbf{a}'\bar{\mathbf{y}} \pm \sqrt{\frac{(n-1)p}{n-p} f_{(1-\alpha, p, n-p)} \frac{\mathbf{a}'\mathbf{S}\mathbf{a}}{n}}.$$

where:

- $\mathbf{a}'\boldsymbol{\mu} = a_1\mu_1 + \cdots + a_p\mu_p$ is a projection onto the axis in the direction of $\mathbf{a}$.
- The probability that at least one interval fails to contain the corresponding $\mathbf{a}'\boldsymbol{\mu}$ is no more than α.
- These intervals are useful for "data snooping" (similar to Scheffé's method in ANOVA 5.1.1.5.4.2).

6.1.6.3.3 One-at-a-Time Confidence Intervals

A simpler alternative is to construct separate confidence intervals for each mean component individually:

$$\bar{y}_i \pm t_{(1-\alpha/2,n-1)}\sqrt{\frac{s_{ii}}{n}}.$$

Limitations

- Each interval has a probability of $1-\alpha$ of covering the corresponding μ_i.
- They ignore the covariance structure between the p variables.

Bonferroni Correction for Multiple Comparisons

If we only care about k specific intervals, we can adjust for multiple comparisons using the Bonferroni correction:

$$\bar{y}_i \pm t_{(1-\alpha/(2k),n-1)}\sqrt{\frac{s_{ii}}{n}}.$$

- This ensures that the overall confidence level remains at $100(1-\alpha)\%$.
- The method becomes more conservative as the number of comparisons k increases.

```
# Load necessary libraries
library(MASS)     # For multivariate analysis
library(ICSNP)    # For Hotelling's T2 test
library(tidyverse)  # Data manipulation and plotting

# Simulated dataset (5 variables, 30 observations)
set.seed(123)
n <- 30  # Sample size
p <- 5   # Number of variables
alpha <- 0.05  # Significance level

# Population mean and covariance
mu <- rep(0, p)
Sigma <- diag(p)

# Generate multivariate normal data
data <- mvrnorm(n, mu, Sigma)
colnames(data) <- paste0("V", 1:p)

# Compute sample mean and covariance
sample_mean <- colMeans(data)
```

```r
sample_cov  <- cov(data)

# Hotelling's T^2 statistic
T2 <-
    n * t(sample_mean - mu) %*% solve(sample_cov) %*%
↪  (sample_mean - mu)

# Critical value for Hotelling's T^2 test
F_crit <- ((n - 1) * p / (n - p)) * qf(1 - alpha, p, n - p)

# Confidence region check
T2 <= F_crit  # If TRUE, mean vector is within the confidence
↪  region
#>      [,1]
#> [1,] TRUE

# Simultaneous confidence intervals
CI_limits <-
    sqrt(((n - 1) * p) / (n - p) * qf(1 - alpha, p, n - p) *
    ↪  diag(sample_cov) / n)

# Construct confidence intervals
simultaneous_CI <- data.frame(
  Variable = colnames(data),
  Lower = sample_mean - CI_limits,
  Upper = sample_mean + CI_limits
)

print(simultaneous_CI)
#>    Variable      Lower     Upper
#> V1       V1 -0.9983080 0.6311472
#> V2       V2 -0.7372215 0.5494437
#> V3       V3 -0.5926088 0.6414496
#> V4       V4 -0.4140990 0.7707756
#> V5       V5 -0.7430441 0.6488366

# Bonferroni-corrected one-at-a-time confidence intervals
t_crit <- qt(1 - alpha / (2 * p), n - 1)

bonferroni_CI <- data.frame(
  Variable = colnames(data),
  Lower = sample_mean - t_crit * sqrt(diag(sample_cov) / n),
  Upper = sample_mean + t_crit * sqrt(diag(sample_cov) / n)
)
```

```
print(bonferroni_CI)
#>    Variable      Lower     Upper
#> V1       V1 -0.7615465 0.3943857
#> V2       V2 -0.5502678 0.3624900
#> V3       V3 -0.4132989 0.4621397
#> V4       V4 -0.2419355 0.5986122
#> V5       V5 -0.5408025 0.4465950
```

6.1.7 General Hypothesis Testing

6.1.7.1 One-Sample Multivariate Tests

We consider testing the hypothesis:

$$H_0 : \mathbf{C}\boldsymbol{\mu} = 0$$

where:

- $\mathbf{C}$ is a $c \times p$ contrast matrix of rank c, where $c \leq p$.
- $\boldsymbol{\mu}$ is the $p \times 1$ population mean vector.

The test statistic for this hypothesis is:

$$F = \frac{n - c}{(n - 1)c} T^2$$

where:

$$T^2 = n(\mathbf{C}\bar{\mathbf{y}})'(\mathbf{C}\mathbf{S}\mathbf{C}')^{-1}(\mathbf{C}\bar{\mathbf{y}}).$$

This follows an F-distribution:

$$F \sim f_{(c, n-c)}.$$

Example: Testing Equal Means Across Variables
We test whether all mean components are equal:

$$H_0 : \mu_1 = \mu_2 = \cdots = \mu_p.$$

This can be rewritten as:

$$\mu_1 - \mu_2 = 0,$$
$$\mu_2 - \mu_3 = 0,$$
$$\vdots$$
$$\mu_{p-1} - \mu_p = 0.$$

Since we are testing $p - 1$ constraints, the contrast matrix $\mathbf{C}$ is a $(p - 1) \times p$ matrix:

$$\mathbf{C} = \begin{bmatrix} 1 & -1 & 0 & \dots & 0 \\ 0 & 1 & -1 & \dots & 0 \\ \vdots & \vdots & \vdots & \ddots & \vdots \\ 0 & 0 & \dots & 1 & -1 \end{bmatrix}.$$

Alternatively, we can compare all other means to the first mean:

$$H_0 : \mu_1 - \mu_2 = 0, \quad \mu_1 - \mu_3 = 0, \quad \dots, \quad \mu_1 - \mu_p = 0.$$

The contrast matrix $\mathbf{C}$ then becomes:

$$\mathbf{C} = \begin{bmatrix} -1 & 1 & 0 & \dots & 0 \\ -1 & 0 & 1 & \dots & 0 \\ \vdots & \vdots & \vdots & \ddots & \vdots \\ -1 & 0 & \dots & 0 & 1 \end{bmatrix}.$$

Key Property

- The value of T^2 is invariant to these different choices of $\mathbf{C}$.

Application: Repeated Measures Design

Repeated measures designs involve measuring each subject multiple times under different conditions or time points.

Let:

- y_{ij} be the response of subject i at time j, where $i = 1, \dots, n$ and $j = 1, \dots, T$.
- $\mathbf{y}_i = (y_{i1}, \dots, y_{iT})'$ be a random sample from:

$$N_T(\boldsymbol{\mu}, \boldsymbol{\Sigma}).$$

Example: Testing Equal Means Over Time
Suppose we have:

- $n = 8$ subjects,
- $T = 6$ time points.

We test:

$$H_0 : \mu_1 = \mu_2 = \cdots = \mu_6.$$

This is equivalent to:

$$\begin{aligned}\mu_1 - \mu_2 &= 0,\\ \mu_2 - \mu_3 &= 0,\\ &\ldots,\\ \mu_5 - \mu_6 &= 0.\end{aligned}$$

The corresponding **contrast matrix** is:

$$\mathbf{C} = \begin{bmatrix} 1 & -1 & 0 & 0 & 0 & 0 \\ 0 & 1 & -1 & 0 & 0 & 0 \\ 0 & 0 & 1 & -1 & 0 & 0 \\ 0 & 0 & 0 & 1 & -1 & 0 \\ 0 & 0 & 0 & 0 & 1 & -1 \end{bmatrix}.$$

If measurements occur at equally spaced time points, we can test for trend effects using orthogonal polynomials.

For example, testing whether quadratic and cubic trends are jointly zero, we use:

$$\mathbf{C} = \begin{bmatrix} 1 & -1 & -1 & 1 \\ -1 & 3 & -3 & 1 \end{bmatrix}.$$

```
# Load necessary libraries
library(MASS)     # For multivariate normal data
library(ICSNP)    # For Hotelling's T^2 test

# Simulated dataset (6 variables, 8 subjects)
set.seed(123)
n <- 8    # Number of subjects
p <- 6    # Number of time points

# Generate sample data
```

```
mu <- rep(5, p)  # Population mean
Sigma <- diag(p)  # Identity covariance matrix

data <- mvrnorm(n, mu, Sigma)
colnames(data) <- paste0("Time", 1:p)

# Compute sample mean and covariance
sample_mean <- colMeans(data)
sample_cov  <- cov(data)

# Define contrast matrix for equal means hypothesis
C <- matrix(0, nrow = p - 1, ncol = p)
for (i in 1:(p - 1)) {
  C[i, i] <- 1
  C[i, i + 1] <- -1
}

# Compute Hotelling's T^2 statistic
T2 <-
    n * t(C %*% sample_mean) %*% solve(C %*% sample_cov %*%
 ↪  t(C)) %*% (C %*% sample_mean)

# Compute F statistic
c <- nrow(C)
F_stat <- ((n - c) / ((n - 1) * c)) * T2

# Critical value
F_crit <- qf(0.95, c, n - c)

# Decision rule
decision <- F_stat > F_crit

# Print results
list(
  T2_statistic = T2,
  F_statistic = F_stat,
  F_critical_value = F_crit,
  Reject_H0 = decision
)
#> $T2_statistic
#>          [,1]
#> [1,] 22.54896
#>
#> $F_statistic
```

```
#>              [,1]
#> [1,] 1.932768
#>
#> $F_critical_value
#> [1] 9.013455
#>
#> $Reject_H0
#>          [,1]
#> [1,] FALSE
```

6.1.7.2 Two-Sample Multivariate Tests

Consider testing the equality of two multivariate population means. Suppose we have two independent random samples:

$$\mathbf{y}_{1i} \sim N_p(\boldsymbol{\mu}_1, \boldsymbol{\Sigma}), \quad i = 1, \dots, n_1,$$
$$\mathbf{y}_{2j} \sim N_p(\boldsymbol{\mu}_2, \boldsymbol{\Sigma}), \quad j = 1, \dots, n_2.$$

We assume:

- Multivariate normality of both populations.
- Equal variance-covariance matrices: $\boldsymbol{\Sigma}_1 = \boldsymbol{\Sigma}_2 = \boldsymbol{\Sigma}$
- Independence between samples

We summarize our data using the sufficient statistics:

- Sample means: $\bar{\mathbf{y}}_1, \bar{\mathbf{y}}_2$
- Sample covariance matrices: $\mathbf{S}_1, \mathbf{S}_2$
- Sample sizes: n_1, n_2

Since we assume equal variance-covariance matrices, we compute a pooled estimator:

$$\mathbf{S} = \frac{(n_1 - 1)\mathbf{S}_1 + (n_2 - 1)\mathbf{S}_2}{(n_1 - 1) + (n_2 - 1)}$$

with $n_1 + n_2 - 2$ degrees of freedom.

We test:

$$H_0 : \boldsymbol{\mu}_1 = \boldsymbol{\mu}_2,$$
$$H_a : \boldsymbol{\mu}_1 \neq \boldsymbol{\mu}_2.$$

That is, we check whether at least one element of $\boldsymbol{\mu}_1 - \boldsymbol{\mu}_2$ is different.

We use:

- $\bar{\mathbf{y}}_1 - \bar{\mathbf{y}}_2$ to estimate $\boldsymbol{\mu}_1 - \boldsymbol{\mu}_2$
- $\mathbf{S}$ to estimate $\boldsymbol{\Sigma}$

Since the two populations are **independent**, the covariance is:

$$\mathrm{Cov}(\bar{\mathbf{y}}_1 - \bar{\mathbf{y}}_2) = \mathrm{Var}(\bar{\mathbf{y}}_1) + \mathrm{Var}(\bar{\mathbf{y}}_2) = \boldsymbol{\Sigma}\left(\frac{1}{n_1} + \frac{1}{n_2}\right).$$

The Hotelling's T^2 statistic is:

$$T^2 = (\bar{\mathbf{y}}_1 - \bar{\mathbf{y}}_2)' \left\{\mathbf{S}\left(\frac{1}{n_1} + \frac{1}{n_2}\right)\right\}^{-1} (\bar{\mathbf{y}}_1 - \bar{\mathbf{y}}_2).$$

which simplifies to:

$$T^2 = \frac{n_1 n_2}{n_1 + n_2}(\bar{\mathbf{y}}_1 - \bar{\mathbf{y}}_2)'\mathbf{S}^{-1}(\bar{\mathbf{y}}_1 - \bar{\mathbf{y}}_2).$$

Reject H_0 if:

$$T^2 \geq \frac{(n_1 + n_2 - 2)p}{n_1 + n_2 - p - 1} f_{(1-\alpha, p, n_1+n_2-p-1)}$$

or equivalently, using the F-statistic:

$$F = \frac{n_1 + n_2 - p - 1}{(n_1 + n_2 - 2)p} T^2.$$

Reject H_0 if:

$$F \geq f_{(1-\alpha, p, n_1+n_2-p-1)}.$$

A $100(1-\alpha)\%$ confidence region for $\boldsymbol{\mu}_1 - \boldsymbol{\mu}_2$ consists of all vectors $\boldsymbol{\delta}$ satisfying:

$$\frac{n_1 n_2}{n_1 + n_2}(\bar{\mathbf{y}}_1 - \bar{\mathbf{y}}_2 - \boldsymbol{\delta})'\mathbf{S}^{-1}(\bar{\mathbf{y}}_1 - \bar{\mathbf{y}}_2 - \boldsymbol{\delta}) \leq \frac{(n_1 + n_2 - 2)p}{n_1 + n_2 - p - 1} f_{(1-\alpha, p, n_1+n_2-p-1)}.$$

For all linear combinations of $\boldsymbol{\mu}_1 - \boldsymbol{\mu}_2$, the simultaneous confidence intervals:

$$\mathbf{a}'(\bar{\mathbf{y}}_1 - \bar{\mathbf{y}}_2) \pm \sqrt{\frac{(n_1 + n_2 - 2)p}{n_1 + n_2 - p - 1} f_{(1-\alpha, p, n_1+n_2-p-1)} \times \mathbf{a}'\mathbf{S}\mathbf{a}\left(\frac{1}{n_1} + \frac{1}{n_2}\right)}.$$

For k **pairwise comparisons**, Bonferroni intervals are:

$$(\bar{y}_{1i} - \bar{y}_{2i}) \pm t_{(1-\alpha/2k, n_1+n_2-2)} \sqrt{\left(\frac{1}{n_1} + \frac{1}{n_2}\right) s_{ii}}.$$

```
# Load necessary libraries
library(MASS)     # For multivariate analysis
library(ICSNP)    # For Hotelling's T^2 test

# Simulated dataset (p = 4 variables, two groups)
set.seed(123)
n1 <- 20   # Sample size for group 1
n2 <- 25   # Sample size for group 2
p <- 4     # Number of variables

# Generate data for both groups
mu1 <- rep(0, p)  # Mean vector for group 1
mu2 <- rep(1, p)  # Mean vector for group 2
Sigma <- diag(p)  # Identity covariance matrix

data1 <- mvrnorm(n1, mu1, Sigma)
data2 <- mvrnorm(n2, mu2, Sigma)

# Compute sample means and covariance matrices
y1_bar <- colMeans(data1)
y2_bar <- colMeans(data2)
S1 <- cov(data1)
S2 <- cov(data2)

# Compute pooled covariance matrix
S_pooled <- ((n1 - 1) * S1 + (n2 - 1) * S2) / (n1 + n2 - 2)

# Compute Hotelling's T^2 statistic
T2 <- (y1_bar - y2_bar) %*% solve(S_pooled * (1/n1 + 1/n2))
↪  %*% (y1_bar - y2_bar)

# Convert to F-statistic
F_stat <- ((n1 + n2 - p - 1) / ((n1 + n2 - 2) * p)) * T2
F_crit <- qf(0.95, p, n1 + n2 - p - 1)

# Decision rule
decision <- F_stat > F_crit
```

```
# Print results
list(
  T2_statistic = T2,
  F_statistic = F_stat,
  F_critical_value = F_crit,
  Reject_H0 = decision
)
#> $T2_statistic
#>          [,1]
#> [1,] 51.90437
#>
#> $F_statistic
#>          [,1]
#> [1,] 12.07078
#>
#> $F_critical_value
#> [1] 2.605975
#>
#> $Reject_H0
#>      [,1]
#> [1,] TRUE
```

6.1.7.3 Model Assumptions in Multivariate Tests

6.1.7.3.1 Effects of Unequal Covariance Matrices

We assume that the two population covariance matrices are equal ($\mathbf{\Sigma}_1 = \mathbf{\Sigma}_2$), but in reality, this assumption may not hold.

Impact on Type I Error and Power

- If $n_1 = n_2$ (large samples), the impact on Type I error rate and power is minimal.
- If $n_1 > n_2$ and eigenvalues of $\mathbf{\Sigma}_1 \mathbf{\Sigma}_2^{-1}$ are less than 1, the Type I error is inflated.
- If $n_1 > n_2$ and some eigenvalues of $\mathbf{\Sigma}_1 \mathbf{\Sigma}_2^{-1}$ are greater than 1, the Type I error is too small, reducing power.

6.1.7.3.2 Effects of Non-normality

Multivariate tests often assume normality, but real-world data may not follow a normal distribution.

Impact on Test Performance

- Two-sample Hotelling's T^2 test is robust to moderate departures from normality if both populations have similar distributions.
- One-sample Hotelling's T^2 test is more sensitive to lack of normality, especially when the distribution is skewed.

Intuition

- A one-sample test depends on the distribution of individual variables, making it more sensitive to normality violations.
- A two-sample test depends on the distribution of differences, which may be less sensitive to non-normality if both groups have similar distributions.

Solutions

1. **Transform the data** (e.g., log or Box-Cox transformation 6.3) to improve normality.
2. **Use large samples** and rely on the [Central Limit Theorem].
3. **Use alternative tests** that do not assume normality:

   - **Wald's Test** (Chi-square-based test), which does not require:

     - Normality
     - Equal sample sizes
     - Equal covariance matrices

   - Test:

$$H_0 : \boldsymbol{\mu}_1 - \boldsymbol{\mu}_2 = 0$$

using:

$$(\bar{\mathbf{y}}_1 - \bar{\mathbf{y}}_2)' \left(\frac{1}{n_1}\mathbf{S}_1 + \frac{1}{n_2}\mathbf{S}_2 \right)^{-1} (\bar{\mathbf{y}}_1 - \bar{\mathbf{y}}_2) \dot{\sim} \chi^2_p.$$

6.1.7.3.3 Testing Equality of Covariance Matrices

With k independent groups, each having a p-dimensional vector, we test:

$$\begin{aligned} H_0 &: \boldsymbol{\Sigma}_1 = \boldsymbol{\Sigma}_2 = \cdots = \boldsymbol{\Sigma}_k = \boldsymbol{\Sigma}, \\ H_a &: \text{At least two are different.} \end{aligned}$$

If H_0 holds, we use a pooled covariance estimate:

$$\mathbf{S} = \frac{\sum_{i=1}^{k}(n_i - 1)\mathbf{S}_i}{\sum_{i=1}^{k}(n_i - 1)}$$

with $\sum_{i=1}^{k}(n_i - 1)$ degrees of freedom.

6.1.7.3.4 Bartlett's Test for Equal Covariances

Bartlett's test is a likelihood ratio test for equality of covariance matrices.

Define:

$$N = \sum_{i=1}^{k} n_i.$$

Compute:

$$M = (N - k) \log |\mathbf{S}| - \sum_{i=1}^{k} (n_i - 1) \log |\mathbf{S}_i|.$$

Correction factor:

$$C^{-1} = 1 - \frac{2p^2 + 3p - 1}{6(p+1)(k-1)} \left\{ \sum_{i=1}^{k} \left(\frac{1}{n_i - 1} \right) - \frac{1}{N-k} \right\}.$$

Reject H_0 if:

$$MC^{-1} > \chi^2_{1-\alpha,(k-1)p(p+1)/2}.$$

Limitations

- Sensitive to non-normality: If data are not normal, MC^{-1} often follows a right-skewed distribution (i.e., shifted to the right of the nominal χ^2 distribution), increasing false positives.
- Best practice: Check univariate and multivariate normality first before using Bartlett's test.

```
# Load required packages
library(MASS)     # For multivariate normal data
library(ICSNP)    # Multivariate tests
library(car)      # Homogeneity of variance tests

# Simulated dataset (three groups, p = 4 variables)
set.seed(123)
n1 <- 20  # Group 1 sample size
n2 <- 25  # Group 2 sample size
n3 <- 30  # Group 3 sample size
p <- 4    # Number of variables
```

```
# Generate data from different covariance structures
mu1 <- rep(0, p)
mu2 <- rep(1, p)
mu3 <- rep(2, p)

Sigma1 <- diag(p)          # Identity covariance for group 1
Sigma2 <- 2 * diag(p)      # Scaled identity for group 2
Sigma3 <- matrix(0.5, p, p) + diag(0.5, p)  # Structured
↪  covariance for group 3

data1 <- mvrnorm(n1, mu1, Sigma1)
data2 <- mvrnorm(n2, mu2, Sigma2)
data3 <- mvrnorm(n3, mu3, Sigma3)

# Create a combined dataset
group_labels <- c(rep("Group1", n1), rep("Group2", n2),
↪  rep("Group3", n3))
data <- data.frame(Group = group_labels, rbind(data1, data2,
↪  data3))

# Compute covariance matrices
S1 <- cov(data1)
S2 <- cov(data2)
S3 <- cov(data3)

# Bartlett's Test for Equal Covariances
bartlett_test <- bartlett.test(data[,-1], g = data$Group)
print(bartlett_test)
#>
#>  Bartlett test of homogeneity of variances
#>
#> data:  data[, -1]
#> Bartlett's K-squared = 0.99333, df = 3, p-value = 0.8029

# Box's M test (alternative for multivariate homogeneity)
box_test <- boxM(data[,-1], data$Group)
print(box_test)
#>
#>  Box's M-test for Homogeneity of Covariance Matrices
#>
#> data:  data[, -1]
#> Chi-Sq (approx.) = 51.039, df = 20, p-value = 0.000157
```

6.1.7.4 Two-Sample Repeated Measures Analysis

Define $\mathbf{y}_{hi}$ as the t-dimensional response vector for subject i in group h:

$$\mathbf{y}_{hi} = (y_{hi1}, y_{hi2}, \dots, y_{hit})'$$

Assume:

- **Group 1**: $\mathbf{y}_{11}, \dots, \mathbf{y}_{1n_1} \sim N_t(\boldsymbol{\mu}_1, \boldsymbol{\Sigma})$ (i.e., iid from a common distribution).
- **Group 2**: $\mathbf{y}_{21}, \dots, \mathbf{y}_{2n_2} \sim N_t(\boldsymbol{\mu}_2, \boldsymbol{\Sigma})$.

We test whether the mean response vectors are **equal across groups**:

$$H_0 : \mathbf{C}(\boldsymbol{\mu}_1 - \boldsymbol{\mu}_2) = \mathbf{0}_c.$$

where:

- $\mathbf{C}$ is a contrast matrix of dimensions $c \times t$ (rank c, where $c \le t$).
- If H_0 is true, the two groups have the same mean structure.

The **Hotelling's T^2** statistic for repeated measures is:

$$T^2 = \frac{n_1 n_2}{n_1 + n_2} (\bar{\mathbf{y}}_1 - \bar{\mathbf{y}}_2)' \mathbf{C}' (\mathbf{CSC}')^{-1} \mathbf{C} (\bar{\mathbf{y}}_1 - \bar{\mathbf{y}}_2).$$

where $\mathbf{S}$ is the pooled covariance matrix. The corresponding F-statistic follows:

$$F = \frac{n_1 + n_2 - c - 1}{(n_1 + n_2 - 2)c} T^2 \sim f_{(c, n_1+n_2-c-1)}.$$

under the null hypothesis.
If we reject $H_0 : \boldsymbol{\mu}_1 = \boldsymbol{\mu}_2$, we may test whether the **profiles are parallel**:

$$\mu_{11} - \mu_{21} = \mu_{12} - \mu_{22},$$
$$\vdots$$
$$\mu_{1t-1} - \mu_{2t-1} = \mu_{1t} - \mu_{2t}.$$

This is expressed as:

$$H_0 : \mathbf{C}(\mu_1 - \mu_2) = \mathbf{0}_c,$$

where:

- $c = t - 1$ (one fewer than the number of time points).

- The contrast matrix **C** is:

$$\mathbf{C} = \begin{bmatrix} 1 & -1 & 0 & \dots & 0 \\ 0 & 1 & -1 & \dots & 0 \\ \vdots & \vdots & \vdots & \ddots & \vdots \\ 0 & 0 & 0 & \dots & -1 \end{bmatrix}_{(t-1)\times t}.$$

1. **One-Sample Hotelling's** T^2 Test

```
# Load necessary libraries
library(ICSNP)
library(dplyr)

# Data: Measurements on 3 variables
plants <- data.frame(
    y1 = c(2.11, 2.36, 2.13, 2.78, 2.17),
    y2 = c(10.1, 35.0, 2.0, 6.0, 2.0),
    y3 = c(3.4, 4.1, 1.9, 3.8, 1.7)
)

# Center the data with hypothesized means
plants_ctr <- plants %>%
    transmute(y1_ctr = y1 - 2.85,
              y2_ctr = y2 - 15.0,
              y3_ctr = y3 - 6.0) %>%
    as.matrix()

# Perform Wilks' Lambda test for one-sample Hotelling's T^2
onesamp_fit <- anova(lm(plants_ctr ~ 1), test = "Wilks")
print(onesamp_fit)
#> Analysis of Variance Table
#>
#>             Df    Wilks approx F num Df den Df  Pr(>F)
#> (Intercept)  1 0.054219   11.629      3      2 0.08022 .
#> Residuals    4
#> ---
#> Signif. codes:  0 '***' 0.001 '**' 0.01 '*' 0.05 '.' 0.1 ' ' 1
```

- If the p-value is large, we fail to reject H_0 and conclude that the hypothesized mean vector is plausible.

- If the p-value is small, we reject H_0 and infer that the sample mean significantly differs from the hypothesized values.

2. **Paired-Sample Hotelling's T^2 Test**

   Used when each subject has two sets of paired measurements.

```
# Data: Commercial vs. State Lab Waste Analysis
waste <- data.frame(
    case = 1:11,
    com_y1 = c(6, 6, 18, 8, 11, 34, 28, 71, 43, 33, 20),
    com_y2 = c(27, 23, 64, 44, 30, 75, 26, 124, 54, 30, 14),
    state_y1 = c(25, 28, 36, 35, 15, 44, 42, 54, 34, 29, 39),
    state_y2 = c(15, 13, 22, 29, 31, 64, 30, 64, 56, 20, 21)
)

# Compute differences between commercial and state labs
waste_diff <- waste %>%
    transmute(y1_diff = com_y1 - state_y1,
              y2_diff = com_y2 - state_y2)

# Perform Paired Hotelling's T^2 test
paired_fit <- HotellingsT2(waste_diff)
print(paired_fit)
#>
#>  Hotelling's one sample T2-test
#>
#> data:  waste_diff
#> T.2 = 6.1377, df1 = 2, df2 = 9, p-value = 0.02083
#> alternative hypothesis: true location is not equal to
↪  c(0,0)
```

- Reject H_0: Measurements from the two labs significantly differ.
- Fail to reject H_0: No significant difference between the two labs.

3. **Independent-Sample Hotelling's T^2 Test with Bartlett's Test**

   Used when comparing two independent groups (Fig. 6.7).

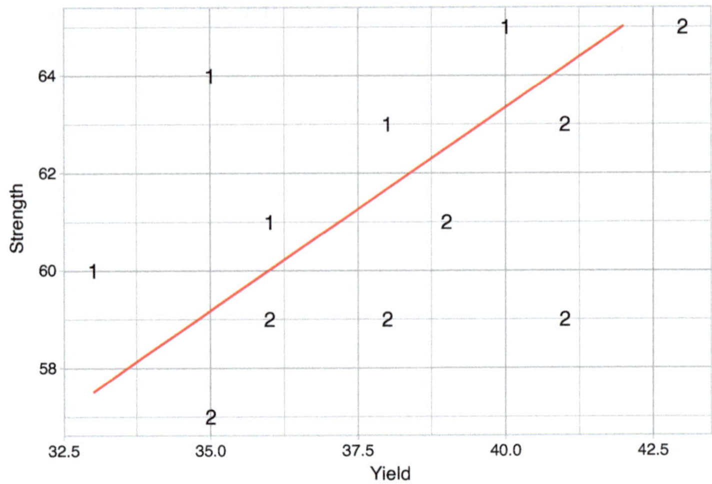

Fig. 6.7 Strength versus yield

```
# Read steel strength data
steel <- read.table("images/steel.dat")
names(steel) <- c("Temp", "Yield", "Strength")

# Scatter plot of Yield vs Strength
library(ggplot2)
```

```
ggplot(steel, aes(x = Yield, y = Strength)) +
    geom_text(aes(label = Temp), size = 5) +
    geom_segment(aes(x = 33, y = 57.5, xend = 42, yend = 65),
    ↪ col = "red")
```

```
# Bartlett's test for equality of covariances
bart_test <- boxM(steel[, -1], steel$Temp)
print(bart_test)  # If p > 0.05, fail to reject equal
↪ covariances
#>
#>  Box's M-test for Homogeneity of Covariance Matrices
#>
#> data:  steel[, -1]
#> Chi-Sq (approx.) = 0.38077, df = 3, p-value = 0.9442

# Multivariate analysis of variance (MANOVA) using Wilks'
↪ Lambda
```

```
twosamp_fit <-
    anova(lm(cbind(Yield, Strength) ~ factor(Temp), data =
    ↪  steel),
          test = "Wilks")
print(twosamp_fit)
#> Analysis of Variance Table
#>
#>                Df    Wilks approx F num Df den Df    Pr(>F)
↪
#> (Intercept)     1 0.001177   3818.1      2      9 6.589e-14
↪   ***
#> factor(Temp)    1 0.294883     10.8      2      9  0.004106
↪   **
#> Residuals      10
↪
#> ---
#> Signif. codes:  0 '***' 0.001 '**' 0.01 '*' 0.05 '.' 0.1 ' ' 1

# Independent-Sample Hotelling's T^2 Test
twosamp_fit2 <-
    HotellingsT2(cbind(steel$Yield, steel$Strength) ~
    ↪  factor(steel$Temp))
print(twosamp_fit2)
#>
#>  Hotelling's two sample T2-test
#>
#> data:  cbind(steel$Yield, steel$Strength) by
↪  factor(steel$Temp)
#> T.2 = 10.76, df1 = 2, df2 = 9, p-value = 0.004106
#> alternative hypothesis: true location difference is not
↪  equal to c(0,0)
```

- Reject H_0: The two temperature groups have significantly different mean vectors.
- Fail to reject H_0: No significant difference between groups.

For a summary of repeated measures hypothesis testing, see Table 6.1.

6.2 Multivariate Analysis of Variance

Multivariate Analysis of Variance (MANOVA) is an extension of the univariate Analysis of Variance (ANOVA) that allows researchers to examine multiple dependent variables simultaneously. Unlike ANOVA, which evaluates differences

Table 6.1 Summary of repeated measures hypothesis testing

Test	Hypothesis	Application
One-Sample Hotelling's T^2	$H_0 : \boldsymbol{\mu} = \boldsymbol{\mu}_0$	Single group mean vector test
Paired-Sample Hotelling's T^2	$H_0 : \boldsymbol{\mu}_d = 0$	Paired measurements comparison
Independent-Sample Hotelling's T^2	$H_0 : \boldsymbol{\mu}_1 = \boldsymbol{\mu}_2$	Two-group mean vector test
Parallel Profiles Test	$H_0 : \mathbf{C}(\boldsymbol{\mu}_1 - \boldsymbol{\mu}_2) = \mathbf{0}$	Testing parallel time trends

in means for a single dependent variable across groups, MANOVA assesses whether there are statistically significant differences among groups across two or more correlated dependent variables.

By considering multiple dependent variables at once, MANOVA accounts for interdependencies between them, reducing the likelihood of Type I errors that may arise from conducting multiple separate ANOVA tests. It is particularly useful in fields such as psychology, marketing, and social sciences, where multiple outcome measures are often interrelated.

This technique is commonly applied in experimental and observational studies where researchers seek to determine the impact of categorical independent variables on multiple continuous dependent variables.

6.2.1 *One-Way MANOVA*

One-way MANOVA extends the univariate one-way ANOVA to multiple dependent variables. It is used to compare treatment means across h different populations when the response consists of multiple correlated variables.

Let the populations be indexed by $i = 1, 2, \ldots, h$, and the observations within each population be indexed by $j = 1, 2, \ldots, n_i$. We assume:

Population 1: $\mathbf{y}_{11}, \mathbf{y}_{12}, \ldots, \mathbf{y}_{1n_1} \sim \text{i.i.d. } N_p(\boldsymbol{\mu}_1, \boldsymbol{\Sigma})$

$\vdots$

Population h: $\mathbf{y}_{h1}, \mathbf{y}_{h2}, \ldots, \mathbf{y}_{hn_h} \sim \text{i.i.d. } N_p(\boldsymbol{\mu}_h, \boldsymbol{\Sigma})$

where:

- $\mathbf{y}_{ij}$ is a p-dimensional response vector for the jth observation in the ith group.
- $\boldsymbol{\mu}_i$ is the population mean vector for the ith group.
- $\boldsymbol{\Sigma}$ is the common covariance matrix across all groups.

Assumptions

1. **Independence**: Observations within and across groups are independent.
2. **Multivariate Normality**: Each population follows a p-variate normal distribution.

3. **Homogeneity of Covariance Matrices**: The covariance matrix $\mathbf{\Sigma}$ is the same for all groups.
   For each group i, we can compute:

- Sample mean vector: $\bar{\mathbf{y}}_i = \frac{1}{n_i}\sum_{j=1}^{n_i}\mathbf{y}_{ij}$
- Sample covariance matrix: $\mathbf{S}_i = \frac{1}{n_i-1}\sum_{j=1}^{n_i}(\mathbf{y}_{ij}-\bar{\mathbf{y}}_i)(\mathbf{y}_{ij}-\bar{\mathbf{y}}_i)'$
- Pooled covariance matrix:

$$\mathbf{S} = \frac{1}{\sum_{i=1}^{h}(n_i-1)}\sum_{i=1}^{h}(n_i-1)\mathbf{S}_i$$

6.2.1.1 Effects Model Formulation

Similar to the univariate one-way ANOVA, the effects model can be written as:

$$\boldsymbol{\mu}_i = \boldsymbol{\mu} + \boldsymbol{\tau}_i$$

where:

- $\boldsymbol{\mu}_i$ is the mean vector for group i.
- $\boldsymbol{\mu}$ is the overall mean effect.
- $\boldsymbol{\tau}_i$ is the treatment effect for group i.

The observational model is:

$$\mathbf{y}_{ij} = \boldsymbol{\mu} + \boldsymbol{\tau}_i + \boldsymbol{\epsilon}_{ij}$$

where $\boldsymbol{\epsilon}_{ij} \sim N_p(\mathbf{0}, \mathbf{\Sigma})$ represents the residual variation.

Since the model is overparameterized, we impose the constraint:

$$\sum_{i=1}^{h} n_i\boldsymbol{\tau}_i = \mathbf{0}$$

or equivalently, we may set $\boldsymbol{\tau}_h = \mathbf{0}$.

Analogous to univariate ANOVA, the total variability is partitioned as:

$$\sum_{i=1}^{h}\sum_{j=1}^{n_i}(\mathbf{y}_{ij}-\bar{\mathbf{y}})(\mathbf{y}_{ij}-\bar{\mathbf{y}})' = \sum_{i=1}^{h} n_i(\bar{\mathbf{y}}_i-\bar{\mathbf{y}})(\bar{\mathbf{y}}_i-\bar{\mathbf{y}})' + \sum_{i=1}^{h}\sum_{j=1}^{n_i}(\mathbf{y}_{ij}-\bar{\mathbf{y}}_i)(\mathbf{y}_{ij}-\bar{\mathbf{y}}_i)'$$

where:

- **LHS**: Total corrected sums of squares and cross-products (SSCP) matrix.
- **RHS**:

Table 6.2 MANOVA partitioning of SSCP and degrees of freedom

Source	SSCP	df
Treatment	$\mathbf{H}$	$h-1$
Residual (error)	$\mathbf{E}$	$\sum_{i=1}^{h} n_i - h$
Total corrected	$\mathbf{H}+\mathbf{E}$	$\sum_{i=1}^{h} n_i - 1$

- First term: between-groups SSCP matrix (denoted **H**)
- Second term: within-groups (residual) SSCP matrix (denoted **E**)

The total within-group variation is (Table 6.2):

$$\mathbf{E} = (n_1 - 1)\mathbf{S}_1 + \cdots + (n_h - 1)\mathbf{S}_h = (\sum_{i=1}^{h} n_i - h)\mathbf{S}$$

6.2.1.1.1 Hypothesis Testing

The null hypothesis states:

$$H_0 : \boldsymbol{\tau}_1 = \boldsymbol{\tau}_2 = \cdots = \boldsymbol{\tau}_h = \mathbf{0}$$

which implies that all group mean vectors are equal.

To test H_0, we assess the relative sizes of $\mathbf{E}$ and $\mathbf{H}+\mathbf{E}$ using Wilks' Lambda: Wilks' Lambda is defined as:

$$\Lambda^* = \frac{|\mathbf{E}|}{|\mathbf{H}+\mathbf{E}|}$$

Properties:

1. In the univariate case, Wilks' Lambda reduces to the F-statistic.
2. The exact distribution of Λ^* is known in special cases.
3. For large samples, we reject H_0 if:

$$-\left(\sum_{i=1}^{h} n_i - 1 - \frac{p+h}{2}\right)\log(\Lambda^*) > \chi^2_{(1-\alpha, p(h-1))}$$

```
# Load dataset
data(iris)

# Fit MANOVA model
manova_fit <-
    manova(cbind(Sepal.Length, Sepal.Width, Petal.Length,
     ↪  Petal.Width) ~ Species,
           data = iris)

# Summary of the MANOVA test
summary(manova_fit, test = "Wilks")
#>            Df    Wilks approx F num Df den Df    Pr(>F)
#> Species     2 0.023439   199.15      8    288 < 2.2e-16 ***
#> Residuals 147
#> ---
#> Signif. codes:  0 '***' 0.001 '**' 0.01 '*' 0.05 '.' 0.1 ' ' 1
```

6.2.1.2 Testing General Hypotheses in MANOVA

We consider h different treatments, where the i-th treatment is applied to n_i subjects, each observed for p repeated measures. This results in a p-dimensional observation vector for each subject in a random sample from each of the h different treatment populations.

The general MANOVA model can be written as:

$$\mathbf{y}_{ij} = \boldsymbol{\mu} + \boldsymbol{\tau}_i + \boldsymbol{\epsilon}_{ij}, \quad i = 1, \dots, h; \quad j = 1, \dots, n_i$$

Equivalently, in matrix notation:

$$\mathbf{Y} = \mathbf{XB} + \boldsymbol{\epsilon}$$

where:

- $\mathbf{Y}_{(n \times p)}$ is the matrix of response variables.
- $\mathbf{X}_{(n \times h)}$ is the design matrix. - $\mathbf{B}_{(h \times p)}$ contains the treatment effects.
- $\boldsymbol{\epsilon}_{(n \times p)}$ is the error matrix.

The response matrix:

$$\mathbf{Y}_{(n \times p)} = \begin{bmatrix} \mathbf{y}'_{11} \\ \vdots \\ \mathbf{y}'_{1n_1} \\ \vdots \\ \mathbf{y}'_{hn_h} \end{bmatrix}$$

The coefficient matrix:

$$\mathbf{B}_{(h \times p)} = \begin{bmatrix} \boldsymbol{\mu}' \\ \boldsymbol{\tau}'_1 \\ \vdots \\ \boldsymbol{\tau}'_{h-1} \end{bmatrix}$$

The error matrix:

$$\boldsymbol{\epsilon}_{(n \times p)} = \begin{bmatrix} \boldsymbol{\epsilon}'_{11} \\ \vdots \\ \boldsymbol{\epsilon}'_{1n_1} \\ \vdots \\ \boldsymbol{\epsilon}'_{hn_h} \end{bmatrix}$$

The design matrix $\mathbf{X}$ encodes the treatment assignments:

$$\mathbf{X}_{(n \times h)} = \begin{bmatrix} 1\,1\,0\ldots 0 \\ \vdots\;\vdots\;\vdots\quad\;\vdots \\ 1\,1\,0\ldots 0 \\ \vdots\;\vdots\;\vdots\ldots\vdots \\ 1\,0\,0\ldots 0 \\ \vdots\;\vdots\;\vdots\quad\;\vdots \\ 1\,0\,0\ldots 0 \end{bmatrix}$$

6.2.1.2.1 Estimation

The least squares estimate of $\mathbf{B}$ is given by:

$$\hat{\mathbf{B}} = (\mathbf{X}'\mathbf{X})^{-1}\mathbf{X}'\mathbf{Y}$$

Since the rows of $\mathbf{Y}$ are independent, we assume:

$$\text{Var}(\mathbf{Y}) = \mathbf{I}_n \otimes \boldsymbol{\Sigma}$$

where $\otimes$ denotes the Kronecker product, resulting in an $np \times np$ covariance matrix.

6.2.1.2.2 Hypothesis Testing

The general hypothesis in MANOVA can be written as:

$$H_0 : \mathbf{LBM} = 0$$

$$H_a : \mathbf{LBM} \neq 0$$

where:

- $\mathbf{L}$ is a $(g \times h)$ matrix of full row rank $(g \leq h)$, specifying comparisons across groups.
- $\mathbf{M}$ is a $(p \times u)$ matrix of full column rank $(u \leq p)$, specifying comparisons across traits.

To evaluate the effect of treatments in the MANOVA framework, we compute the treatment corrected sums of squares and cross-product (SSCP) matrix:

$$\mathbf{H} = \mathbf{M'Y'X(X'X)^{-1}L'[L(X'X)^{-1}L']^{-1}L(X'X)^{-1}X'YM}$$

If we are testing the null hypothesis:

$$H_0 : \mathbf{LBM} = \mathbf{D}$$

then the corresponding SSCP matrix is:

$$\mathbf{H} = (\mathbf{L\hat{B}M} - \mathbf{D})'[\mathbf{X(X'X)^{-1}L}]^{-1}(\mathbf{L\hat{B}M} - \mathbf{D})$$

Similarly, the residual (error) SSCP matrix is given by:

$$\mathbf{E} = \mathbf{M'Y'[I - X(X'X)^{-1}X']YM}$$

which can also be expressed as:

$$\mathbf{E} = \mathbf{M'[Y'Y - \hat{B}'(X'X)^{-1}\hat{B}]M}$$

These matrices, $\mathbf{H}$ and $\mathbf{E}$, serve as the basis for assessing the relative treatment effect in a multivariate setting.

6.2.1.2.3 Test Statistics in MANOVA

To test whether treatment effects significantly impact the multivariate response, we examine the eigenvalues of $\mathbf{HE}^{-1}$, leading to several common test statistics:

1. **Wilks' Lambda**:

$$\Lambda^* = \frac{|\mathbf{E}|}{|\mathbf{H} + \mathbf{E}|}$$

A smaller Λ^* indicates a greater difference among group mean vectors. The degrees of freedom depend on the ranks of **L**, **M**, and **X**.

2. **Lawley-Hotelling Trace**:

$$U = \text{tr}(\mathbf{H}\mathbf{E}^{-1})$$

This statistic sums the eigenvalues of $\mathbf{H}\mathbf{E}^{-1}$, capturing the overall treatment effect.

3. **Pillai's Trace**:

$$V = \text{tr}(\mathbf{H}(\mathbf{H} + \mathbf{E})^{-1})$$

This test is known for its robustness against violations of MANOVA assumptions.

4. **Roy's Maximum Root**: The largest eigenvalue of $\mathbf{H}\mathbf{E}^{-1}$. It focuses on the strongest treatment effect present in the data.

For large n, under H_0:

$$-\left(n - 1 - \frac{p+h}{2}\right) \ln \Lambda^* \sim \chi^2_{p(h-1)}$$

In certain cases, specific values of p and h allow for an exact F-distribution under H_0.

```
## One-Way MANOVA

library(car)
library(emmeans)
library(profileR)
library(tidyverse)

# Read in the data
gpagmat <- read.table("images/gpagmat.dat")

# Change the variable names
names(gpagmat) <- c("y1", "y2", "admit")

# Check the structure of the dataset
str(gpagmat)
#> 'data.frame':    85 obs. of  3 variables:
```

```
#>  $ y1   : num  2.96 3.14 3.22 3.29 3.69 3.46 3.03 3.19 3.63
↪  3.59 ...
#>  $ y2   : int  596 473 482 527 505 693 626 663 447 588 ...
#>  $ admit: int  1 1 1 1 1 1 1 1 1 1 ...

# Plot the data
gg <- ggplot(gpagmat, aes(x = y1, y = y2)) +
    geom_text(aes(label = admit, col = as.character(admit))) +
    scale_color_discrete(name = "Admission",
                         labels = c("Admit", "Do not admit",
                         ↪  "Borderline")) +
    scale_x_continuous(name = "GPA") +
    scale_y_continuous(name = "GMAT")

# Fit a one-way MANOVA model
oneway_fit <- manova(cbind(y1, y2) ~ admit, data = gpagmat)

# MANOVA test using Wilks' Lambda
summary(oneway_fit, test = "Wilks")
#>           Df  Wilks approx F num Df den Df    Pr(>F)
#> admit      1 0.6126   25.927      2     82 1.881e-09 ***
#> Residuals 83
#> ---
#> Signif. codes:  0 '***' 0.001 '**' 0.01 '*' 0.05 '.' 0.1 ' ' 1
```

Since the Wilks' Lambda test results in a small p-value, we reject the null hypothesis of equal multivariate mean vectors among the three admission groups.

Repeated Measures MANOVA

```
# Create dataset for repeated measures example
stress <- data.frame(
    subject = 1:8,
    begin = c(3, 2, 5, 6, 1, 5, 1, 5),
    middle = c(3, 4, 3, 7, 4, 7, 1, 2),
    final = c(6, 7, 4, 7, 6, 7, 3, 5)
)
```

Choosing the Correct Model

- If time (with three levels) is treated as an independent variable, we use univariate ANOVA (which requires the sphericity assumption, meaning the variances of all differences must be equal).
- If each time point is treated as a separate variable, we use MANOVA (which does not require the sphericity assumption).

```
# Fit the MANOVA model for repeated measures
stress_mod <- lm(cbind(begin, middle, final) ~ 1, data =
↪ stress)

# Define the within-subject factor
idata <- data.frame(time = factor(
    c("begin", "middle", "final"),
    levels = c("begin", "middle", "final")
))

# Perform repeated measures MANOVA
repeat_fit <- Anova(
    stress_mod,
    idata = idata,
    idesign = ~ time,
    icontrasts = "contr.poly"
)

# Summarize results
summary(repeat_fit)
#>
#> Type III Repeated Measures MANOVA Tests:
#>
#> ------------------------------------------
#>
#> Term: (Intercept)
#>
#>  Response transformation matrix:
#>        (Intercept)
#> begin            1
#> middle           1
#> final            1
#>
#> Sum of squares and products for the hypothesis:
#>             (Intercept)
#> (Intercept)        1352
#>
#> Multivariate Tests: (Intercept)
#>                  Df test stat approx F num Df den Df
↪  Pr(>F)
#> Pillai            1  0.896552 60.66667      1      7
↪  0.00010808 ***
#> Wilks             1  0.103448 60.66667      1      7
↪  0.00010808 ***
```

```
#> Hotelling-Lawley  1  8.666667 60.66667      1      7
↪  0.00010808 ***
#> Roy               1  8.666667 60.66667      1      7
↪  0.00010808 ***
#> ---
#> Signif. codes:  0 '***' 0.001 '**' 0.01 '*' 0.05 '.' 0.1 ' ' 1
#>
#> ------------------------------------------
#>
#> Term: time
#>
#>  Response transformation matrix:
#>               time.L     time.Q
#> begin  -7.071068e-01  0.4082483
#> middle -7.850462e-17 -0.8164966
#> final   7.071068e-01  0.4082483
#>
#> Sum of squares and products for the hypothesis:
#>           time.L   time.Q
#> time.L 18.062500 6.747781
#> time.Q  6.747781 2.520833
#>
#> Multivariate Tests: time
#>                  Df test stat approx F num Df den Df
↪  Pr(>F)
#> Pillai            1 0.7080717 7.276498      2      6
↪  0.024879 *
#> Wilks             1 0.2919283 7.276498      2      6
↪  0.024879 *
#> Hotelling-Lawley  1 2.4254992 7.276498      2      6
↪  0.024879 *
#> Roy               1 2.4254992 7.276498      2      6
↪  0.024879 *
#> ---
#> Signif. codes:  0 '***' 0.001 '**' 0.01 '*' 0.05 '.' 0.1 ' ' 1
#>
#> Univariate Type III Repeated-Measures ANOVA Assuming
↪  Sphericity
#>
#>             Sum Sq num Df Error SS den Df F value    Pr(>F)
↪
#> (Intercept) 450.67      1    52.00      7 60.6667 0.0001081
↪  ***
```

```
#> time           20.58      2    24.75      14  5.8215 0.0144578
↪  *
#> ---
#> Signif. codes:  0 '***' 0.001 '**' 0.01 '*' 0.05 '.' 0.1 ' ' 1
#>
#>
#> Mauchly Tests for Sphericity
#>
#>      Test statistic p-value
#> time         0.7085 0.35565
#>
#>
#> Greenhouse-Geisser and Huynh-Feldt Corrections
#>  for Departure from Sphericity
#>
#>       GG eps Pr(>F[GG])
#> time 0.77429    0.02439 *
#> ---
#> Signif. codes:  0 '***' 0.001 '**' 0.01 '*' 0.05 '.' 0.1 ' ' 1
#>
#>         HF eps Pr(>F[HF])
#> time 0.9528433 0.01611634
```

The results indicate that we cannot reject the null hypothesis of sphericity, meaning that univariate ANOVA is also appropriate. The linear time effect is significant, but the quadratic time effect is not.

Polynomial Contrasts for Time Effects
To further explore the effect of time, we examine polynomial contrasts.

```
# Check the reference for the marginal means
ref_grid(stress_mod, mult.name = "time")
#> 'emmGrid' object with variables:
#>     1 = 1
#>     time = multivariate response levels: begin, middle,
↪  final

# Compute marginal means for time levels
contr_means <- emmeans(stress_mod, ~ time, mult.name = "time")

# Test for polynomial trends
contrast(contr_means, method = "poly")
#>  contrast  estimate    SE df t.ratio p.value
#>  linear        2.12 0.766  7   2.773  0.0276
#>  quadratic     1.38 0.944  7   1.457  0.1885
```

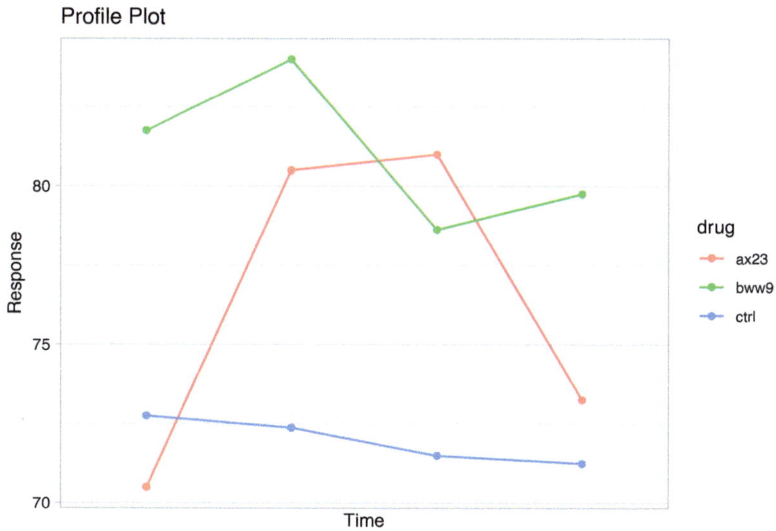

Fig. 6.8 Profile plot

The results confirm that there is a significant linear trend over time but no quadratic trend.

MANOVA for Drug Treatments

We now analyze a multivariate response for different drug treatments (Fig. 6.8).

```
# Read in the dataset
heart <- read.table("images/heart.dat")

# Assign variable names
names(heart) <- c("drug", "y1", "y2", "y3", "y4")

# Create a subject ID nested within drug groups
heart <- heart %>%
    group_by(drug) %>%
    mutate(subject = row_number()) %>%
    ungroup()

# Check dataset structure
str(heart)
#> tibble [24 x 6] (S3: tbl_df/tbl/data.frame)
#>  $ drug    : chr [1:24] "ax23" "ax23" "ax23" "ax23" ...
#>  $ y1      : int [1:24] 72 78 71 72 66 74 62 69 85 82 ...
```

```
#>  $ y2     : int [1:24] 86 83 82 83 79 83 73 75 86 86 ...
#>  $ y3     : int [1:24] 81 88 81 83 77 84 78 76 83 80 ...
#>  $ y4     : int [1:24] 77 82 75 69 66 77 70 70 80 84 ...
#>  $ subject: int [1:24] 1 2 3 4 5 6 7 8 1 2 ...

# Create means summary for a profile plot
heart_means <- heart %>%
    group_by(drug) %>%
    summarize_at(vars(starts_with("y")), mean) %>%
    ungroup() %>%
    pivot_longer(-drug, names_to = "time", values_to = "mean")
    ↪ %>%
    mutate(time = as.numeric(as.factor(time)))

# Generate the profile plot
gg_profile <- ggplot(heart_means, aes(x = time, y = mean)) +
    geom_line(aes(col = drug)) +
    geom_point(aes(col = drug)) +
    ggtitle("Profile Plot") +
    scale_y_continuous(name = "Response") +
    scale_x_discrete(name = "Time")
```

```
gg_profile
```

```
# Fit the MANOVA model
heart_mod <- lm(cbind(y1, y2, y3, y4) ~ drug, data = heart)

# Perform MANOVA test
man_fit <- car::Anova(heart_mod)

# Summarize results
summary(man_fit)
#>
#> Type II MANOVA Tests:
#>
#> Sum of squares and products for error:
#>        y1      y2      y3     y4
#> y1 641.00 601.750 535.250 426.00
#> y2 601.75 823.875 615.500 534.25
#> y3 535.25 615.500 655.875 555.25
```

```
#> y4 426.00 534.250 555.250 674.50
#>
#> ------------------------------------------
#>
#> Term: drug
#>
#> Sum of squares and products for the hypothesis:
#>         y1       y2       y3    y4
#> y1 567.00 335.2500  42.7500 387.0
#> y2 335.25 569.0833 404.5417 367.5
#> y3  42.75 404.5417 391.0833 171.0
#> y4 387.00 367.5000 171.0000 316.0
#>
#> Multivariate Tests: drug
#>                  Df test stat  approx F num Df den Df
↪   Pr(>F)
#> Pillai            2  1.283456  8.508082      8     38
↪   1.5010e-06 ***
#> Wilks             2  0.079007 11.509581      8     36
↪   6.3081e-08 ***
#> Hotelling-Lawley  2  7.069384 15.022441      8     34
↪   3.9048e-09 ***
#> Roy               2  6.346509 30.145916      4     19
↪   5.4493e-08 ***
#> ---
#> Signif. codes:  0 '***' 0.001 '**' 0.01 '*' 0.05 '.' 0.1 ' ' 1
```

Since we obtain a small p-value, we reject the null hypothesis of no difference in means between the treatments.

Contrasts Between Treatment Groups

To further investigate group differences, we define **contrast matrices.**

```
# Convert drug variable to a factor
heart$drug <- factor(heart$drug)

# Define contrast matrix L
L <- matrix(c(0, 2,
              1, -1,-1, -1), nrow = 3, byrow = TRUE)

colnames(L) <- c("bww9:ctrl", "ax23:rest")
rownames(L) <- unique(heart$drug)
```

```r
# Assign contrasts
contrasts(heart$drug) <- L
contrasts(heart$drug)
#>       bww9:ctrl ax23:rest
#> ax23          0         2
#> bww9          1        -1
#> ctrl         -1        -1
```

Hypothesis Testing with Contrast Matrices
Instead of setting contrasts in `heart$drug`, we use a contrast matrix **M**.

```r
# Define contrast matrix M for further testing
M <- matrix(c(1, -1, 0, 0,
              0, 1, -1, 0,
              0, 0, 1, -1), nrow = 4)

# Update model for contrast testing
heart_mod2 <- update(heart_mod)

# Display model coefficients
coef(heart_mod2)
#>                  y1        y2        y3    y4
#> (Intercept)   75.00 78.9583333 77.041667 74.75
#> drugbww9:ctrl  4.50  5.8125000  3.562500  4.25
#> drugax23:rest -2.25  0.7708333  1.979167 -0.75
```

Comparing Drug bww9 vs Control

```r
bww9vctrl <-
    car::linearHypothesis(heart_mod2,
                          hypothesis.matrix = c(0, 1, 0),
                          P = M)
bww9vctrl
#>
#>  Response transformation matrix:
#>    [,1] [,2] [,3]
#> y1    1    0    0
#> y2   -1    1    0
#> y3    0   -1    1
```

```
#> y4      0     0    -1
#>
#> Sum of squares and products for the hypothesis:
#>             [,1]    [,2]      [,3]
#> [1,]  27.5625 -47.25  14.4375
#> [2,] -47.2500  81.00 -24.7500
#> [3,]  14.4375 -24.75   7.5625
#>
#> Sum of squares and products for error:
#>             [,1]      [,2]     [,3]
#> [1,]  261.375 -141.875  28.000
#> [2,] -141.875  248.750 -19.375
#> [3,]   28.000  -19.375 219.875
#>
#> Multivariate Tests:
#>                  Df test stat approx F num Df den Df Pr(>F)
#> Pillai            1 0.2564306 2.184141      3     19 0.1233
#> Wilks             1 0.7435694 2.184141      3     19 0.1233
#> Hotelling-Lawley  1 0.3448644 2.184141      3     19 0.1233
#> Roy               1 0.3448644 2.184141      3     19 0.1233

bww9vctrl <-
    car::linearHypothesis(heart_mod,
                          hypothesis.matrix = c(0, 1,-1),
                          P = M)
bww9vctrl
#>
#>  Response transformation matrix:
#>    [,1] [,2] [,3]
#> y1    1    0    0
#> y2   -1    1    0
#> y3    0   -1    1
#> y4    0    0   -1
#>
#> Sum of squares and products for the hypothesis:
#>             [,1]    [,2]      [,3]
#> [1,]  27.5625 -47.25  14.4375
#> [2,] -47.2500  81.00 -24.7500
#> [3,]  14.4375 -24.75   7.5625
#>
#> Sum of squares and products for error:
#>             [,1]      [,2]     [,3]
#> [1,]  261.375 -141.875  28.000
#> [2,] -141.875  248.750 -19.375
```

```
#> [3,]    28.000  -19.375 219.875
#>
#> Multivariate Tests:
#>                  Df test stat approx F num Df den Df Pr(>F)
#> Pillai            1 0.2564306 2.184141      3     19 0.1233
#> Wilks             1 0.7435694 2.184141      3     19 0.1233
#> Hotelling-Lawley  1 0.3448644 2.184141      3     19 0.1233
#> Roy               1 0.3448644 2.184141      3     19 0.1233
```

Since the p-value is not significant, we conclude that there is no significant difference between the control and bww9 drug treatment.

Comparing Drug ax23 vs Rest

```
axx23vrest <-
    car::linearHypothesis(heart_mod2,
                          hypothesis.matrix = c(0, 0, 1),
                          P = M)
axx23vrest
#>
#>  Response transformation matrix:
#>      [,1] [,2] [,3]
#> y1      1    0    0
#> y2     -1    1    0
#> y3      0   -1    1
#> y4      0    0   -1
#>
#> Sum of squares and products for the hypothesis:
#>           [,1]       [,2]      [,3]
#> [1,]  438.0208  175.20833 -395.7292
#> [2,]  175.2083   70.08333 -158.2917
#> [3,] -395.7292 -158.29167  357.5208
#>
#> Sum of squares and products for error:
#>          [,1]     [,2]    [,3]
#> [1,]  261.375 -141.875  28.000
#> [2,] -141.875  248.750 -19.375
#> [3,]   28.000  -19.375 219.875
#>
#> Multivariate Tests:
#>                  Df test stat approx F num Df den Df
↪   Pr(>F)
```

```
#> Pillai            1  0.855364 37.45483      3      19
↪  3.5484e-08 ***
#> Wilks             1  0.144636 37.45483      3      19
↪  3.5484e-08 ***
#> Hotelling-Lawley  1  5.913921 37.45483      3      19
↪  3.5484e-08 ***
#> Roy               1  5.913921 37.45483      3      19
↪  3.5484e-08 ***
#> ---
#> Signif. codes:  0 '***' 0.001 '**' 0.01 '*' 0.05 '.' 0.1 ' ' 1
```

```
axx23vrest <-
    car::linearHypothesis(heart_mod,
                          hypothesis.matrix = c(2,-1, 1),
                          P = M)
axx23vrest
#>
#>  Response transformation matrix:
#>     [,1] [,2] [,3]
#> y1     1    0    0
#> y2    -1    1    0
#> y3     0   -1    1
#> y4     0    0   -1
#>
#> Sum of squares and products for the hypothesis:
#>            [,1]       [,2]      [,3]
#> [1,]  402.5208  127.41667 -390.9375
#> [2,]  127.4167   40.33333 -123.7500
#> [3,] -390.9375 -123.75000  379.6875
#>
#> Sum of squares and products for error:
#>           [,1]     [,2]    [,3]
#> [1,]  261.375 -141.875  28.000
#> [2,] -141.875  248.750 -19.375
#> [3,]   28.000  -19.375 219.875
#>
#> Multivariate Tests:
#>                  Df test stat approx F num Df den Df
↪  Pr(>F)
#> Pillai            1  0.842450 33.86563      3      19
↪  7.9422e-08 ***
#> Wilks             1  0.157550 33.86563      3      19
↪  7.9422e-08 ***
```

```
#> Hotelling-Lawley  1  5.347205 33.86563      3      19
↪   7.9422e-08 ***
#> Roy               1  5.347205 33.86563      3      19
↪   7.9422e-08 ***
#> ---
#> Signif. codes:  0 '***' 0.001 '**' 0.01 '*' 0.05 '.' 0.1 ' ' 1
```

Since we obtain a significant p-value, we conclude that the `ax23` drug treatment significantly differs from the rest of the treatments.

6.2.2 Profile Analysis

Profile analysis is a multivariate extension of repeated measures ANOVA used to examine similarities between treatment effects in longitudinal data. The null hypothesis states that all treatments have the same average effect:

$$H_0 : \mu_1 = \mu_2 = \cdots = \mu_h$$

which is equivalent to testing:

$$H_0 : \tau_1 = \tau_2 = \cdots = \tau_h$$

This analysis helps determine the exact nature of similarities and differences between treatments by sequentially testing the following hypotheses:

1. Are the profiles parallel? No interaction between treatment and time.
2. Are the profiles coincidental? Identical profiles across treatments.
3. Are the profiles horizontal? No differences across any time points.

If the **parallelism assumption is rejected**, we can further investigate:

- Differences among groups at specific time points.
- Differences among time points within a particular group.
- Differences within subsets of time points across groups.

Example: Profile Analysis Setup
Consider an example with:

- 4 time points ($p = 4$)
- 3 treatments ($h = 3$)

We analyze whether the profiles for each treatment group are identical except for a mean shift (i.e., they are parallel).

6.2.2.1 Parallel Profile Hypothesis

We test whether the differences in treatment means remain constant across time:

$$\begin{aligned} H_0: \mu_{11} - \mu_{21} &= \mu_{12} - \mu_{22} = \cdots = \mu_{1t} - \mu_{2t} \\ \mu_{11} - \mu_{31} &= \mu_{12} - \mu_{32} = \cdots = \mu_{1t} - \mu_{3t} \\ &\vdots \end{aligned}$$

for $h - 1$ equations. Equivalently, in matrix form:

$$H_0 : \mathbf{LBM} = 0$$

where the cell means parameterization of $\mathbf{B}$ is:

$$\mathbf{LBM} = \begin{bmatrix} 1 & -1 & 0 \\ 1 & 0 & -1 \end{bmatrix} \begin{bmatrix} \mu_{11} & \dots & \mu_{14} \\ \mu_{21} & \dots & \mu_{24} \\ \mu_{31} & \dots & \mu_{34} \end{bmatrix} \begin{bmatrix} 1 & 1 & 1 \\ -1 & 0 & 0 \\ 0 & -1 & 0 \\ 0 & 0 & -1 \end{bmatrix} = \mathbf{0}$$

where:

- The first matrix compares treatments at each time point.
- The second matrix contains the mean responses for each treatment across time.
- The third matrix compares time points within each treatment.

The first multiplication, $\mathbf{LB}$, results in:

$$\begin{bmatrix} \mu_{11} - \mu_{21} & \mu_{12} - \mu_{22} & \mu_{13} - \mu_{23} & \mu_{14} - \mu_{24} \\ \mu_{11} - \mu_{31} & \mu_{12} - \mu_{32} & \mu_{13} - \mu_{33} & \mu_{14} - \mu_{34} \end{bmatrix}$$

This represents differences in treatment means at each time point. Multiplying by $\mathbf{M}$ compares these differences across time:

$$\begin{bmatrix} (\mu_{11} - \mu_{21}) - (\mu_{12} - \mu_{22}) & (\mu_{11} - \mu_{21}) - (\mu_{13} - \mu_{23}) & (\mu_{11} - \mu_{21}) - (\mu_{14} - \mu_{24}) \\ (\mu_{11} - \mu_{31}) - (\mu_{12} - \mu_{32}) & (\mu_{11} - \mu_{31}) - (\mu_{13} - \mu_{33}) & (\mu_{11} - \mu_{31}) - (\mu_{14} - \mu_{34}) \end{bmatrix}$$

This matrix captures the changes in treatment differences over time.

Instead of using the cell means parameterization, we can express the model in terms of effects:

$$\mathbf{LBM} = \begin{bmatrix} 0 & 1 & -1 & 0 \\ 0 & 1 & 0 & -1 \end{bmatrix} \begin{bmatrix} \mu' \\ \tau_1' \\ \tau_2' \\ \tau_3' \end{bmatrix} \begin{bmatrix} 1 & 1 & 1 \\ -1 & 0 & 0 \\ 0 & -1 & 0 \\ 0 & 0 & -1 \end{bmatrix} = \mathbf{0}$$

where:

- The first matrix compares treatment effects.
- The second matrix contains overall mean and treatment effects.
- The third matrix compares time points within each treatment.

In both parameterizations:

- $\text{rank}(\mathbf{L}) = h - 1$
- $\text{rank}(\mathbf{M}) = p - 1$

indicating that we are testing for differences across treatments (**L**) and time points (**M**).

Different choices of **L** and **M** still lead to the same hypothesis:

$$\mathbf{L} = \begin{bmatrix} 0 & 1 & 0 & -1 \\ 0 & 0 & 1 & -1 \end{bmatrix}$$

$$\mathbf{M} = \begin{bmatrix} 1 & 0 & 0 \\ -1 & 1 & 0 \\ 0 & -1 & 1 \\ 0 & 0 & -1 \end{bmatrix}$$

This choice yields the same profile comparison structure.

6.2.2.2 Coincidental Profiles

Once we establish that the profiles are **parallel** (i.e., we fail to reject the parallel profile test), we next ask:

Are the Profiles Identical?

If the sums of the components of $\boldsymbol{\mu}_i$ are equal across all treatments, then the profiles are **coincidental** (i.e., identical across groups).

Hypothesis for Coincidental Profiles

$$H_0 : \mathbf{1}'_p \boldsymbol{\mu}_1 = \mathbf{1}'_p \boldsymbol{\mu}_2 = \cdots = \mathbf{1}'_p \boldsymbol{\mu}_h$$

Equivalently, in matrix notation:

$$H_0 : \mathbf{LBM} = \mathbf{0}$$

where, under the **cell means parameterization**:

$$\mathbf{L} = \begin{bmatrix} 1 & 0 & -1 \\ 0 & 1 & -1 \end{bmatrix}$$

and

$$\mathbf{M} = \begin{bmatrix} 1 & 1 & 1 & 1 \end{bmatrix}'$$

Multiplying **LBM** yields:

$$\begin{bmatrix} (\mu_{11} + \mu_{12} + \mu_{13} + \mu_{14}) - (\mu_{31} + \mu_{32} + \mu_{33} + \mu_{34}) \\ (\mu_{21} + \mu_{22} + \mu_{23} + \mu_{24}) - (\mu_{31} + \mu_{32} + \mu_{33} + \mu_{34}) \end{bmatrix} = \begin{bmatrix} 0 \\ 0 \end{bmatrix}$$

Thus, rejecting this hypothesis suggests that the treatments do not have identical effects over time.

As in previous cases, different choices of **L** and **M** can yield the same result.

6.2.2.3 Horizontal Profiles

If we fail to reject the null hypothesis that all h profiles are the same (i.e., we confirm parallel and coincidental profiles), we then ask:

Are All Elements of the Common Profile Equal Across Time?
This question tests whether the profiles are horizontal, meaning no differences between any time points.

Hypothesis for Horizontal Profiles

$$H_0 : \mathbf{LBM} = \mathbf{0}$$

where

$$\mathbf{L} = \begin{bmatrix} 1 & 0 & 0 \end{bmatrix}$$

Table 6.3 Profile tests and their ANOVA equivalents in repeated measures Designsknit

Test	Equivalent test for
Parallel profiles	Interaction effect (treatment × time)
Coincidental profiles	Main effect of treatment (between-subjects factor)
Horizontal profiles	Main effect of time (repeated measures factor)

and

$$\mathbf{M} = \begin{bmatrix} 1 & 0 & 0 \\ -1 & 1 & 0 \\ 0 & -1 & 1 \\ 0 & 0 & -1 \end{bmatrix}$$

Multiplication yields:

$$\left[(\mu_{11} - \mu_{12})\ (\mu_{12} - \mu_{13})\ (\mu_{13} - \mu_{14})\right] = \left[0\ 0\ 0\right]$$

Thus, rejecting this hypothesis suggests that at least one time point differs significantly from the others in the common profile.

6.2.2.4 Summary of Profile Tests

If we fail to reject all three hypotheses, we conclude that:

- There are no significant differences between treatments.
- There are no significant differences between time points.

The three profile tests correspond to well-known univariate ANOVA components (Table 6.3).

```
# Profile analysis for heart dataset
profile_fit <- pbg(
    data = as.matrix(heart[, 2:5]),      # Response variables
    group = as.matrix(heart[, 1]),       # Treatment group
    original.names = TRUE,               # Keep original variable
    ↪ names
    profile.plot = FALSE                 # Disable automatic
    ↪ plotting
)

# Summary of profile analysis results
summary(profile_fit)
```

```
#> Call:
#> pbg(data = as.matrix(heart[, 2:5]), group =
↪   as.matrix(heart[,
#>     1]), original.names = TRUE, profile.plot = FALSE)
#>
#> Hypothesis Tests:
#> $`Ho: Profiles are parallel`
#>   Multivariate.Test Statistic  Approx.F num.df den.df
↪   p.value
#> 1             Wilks 0.1102861 12.737599      6     38
↪   7.891497e-08
#> 2            Pillai 1.0891707  7.972007      6     40
↪   1.092397e-05
#> 3  Hotelling-Lawley 6.2587852 18.776356      6     36
↪   9.258571e-10
#> 4               Roy 5.9550887 39.700592      3     20
↪   1.302458e-08
#>
#> $`Ho: Profiles have equal levels`
#>             Df Sum Sq Mean Sq F value  Pr(>F)
#> group        2  328.7  164.35   5.918 0.00915 **
#> Residuals   21  583.2   27.77
#> ---
#> Signif. codes:  0 '***' 0.001 '**' 0.01 '*' 0.05 '.' 0.1 ' ' 1
#>
#> $`Ho: Profiles are flat`
#>          F df1 df2      p-value
#> 1 14.30928   3  19 4.096803e-05
```

6.3 Statistical Test Selection for Comparing Means

See Fig. 6.9.

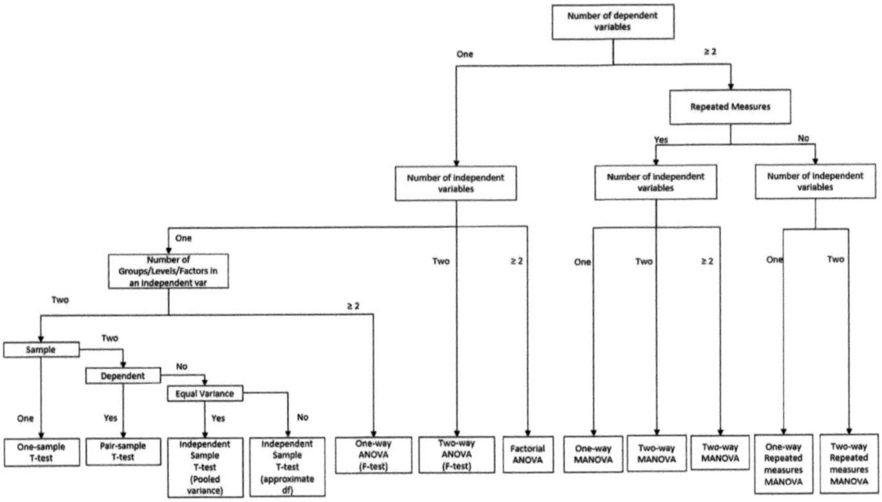

Fig. 6.9 Statistical test selection for comparing means

Bibliography

Abadie, A., Angrist, J., & Imbens, G. (2002). Instrumental variables estimates of the effect of subsidized training on the quantiles of trainee earnings. *Econometrica, 70*(1), 91–117.

Bertrand, M., & Mullainathan, S. (2004). Are emily and greg more employable than lakisha and jamal? a field experiment on labor market discrimination. *American Economic Review, 94*(4), 991–1013.

Braun, M., & Schwartz, E. M. (2025). Where a/b testing goes wrong: How divergent delivery affects what online experiments cannot (and can) tell you about how customers respond to advertising. *Journal of Marketing, 89*, 71–95.

Chen, J., & Roth, J. (2024). Logs with zeros? Some problems and solutions. *The Quarterly Journal of Economics, 139*(2), 891–936.

Chernozhukov, V., & Hansen, C. (2005). An iv model of quantile treatment effects. *Econometrica, 73*(1), 245–261.

Cullen, J. B., Jacob, B. A., & Levitt, S. D. (2005). The impact of school choice on student outcomes: An analysis of the chicago public schools. *Journal of Public Economics, 89*(5–6), 729–760.

Doornik, J. A., & Hansen, H. (2008). An omnibus test for univariate and multivariate normality. *Oxford Bulletin of Economics and Statistics, 70*, 927–939.

Freedman, D. A. (2008). Randomization does not justify logistic regression. *Statistical Science, 23*, 237–249.

Gerber, A. S., Green, D. P., Kaplan, E. H., & Kern, H. L. (2010). Baseline, placebo, and treatment: Efficient estimation for three-group experiments. *Political Analysis, 18*(3), 297–315.

Greevy, R., Lu, B., Silber, J. H., & Rosenbaum, P. (2004). Optimal multivariate matching before randomization. *Biostatistics, 5*(2), 263–275.

Hoyer, P., Janzing, D., Mooij, J. M., Peters, J., & Schölkopf, B. (2008). Nonlinear causal discovery with additive noise models. In *Advances in neural information processing systems*, 21.

Imai, K., Jiang, Z., & Malani, A. (2021). Causal inference with interference and noncompliance in two-stage randomized experiments. *Journal of the American Statistical Association, 116*(534), 632–644.

Imai, K., King, G., & Stuart, E. A. (2008). Misunderstandings between experimentalists and observationalists about causal inference. *Journal of the Royal Statistical Society: Series A (Statistics in Society), 171*(2), 481–502.

Imbens, G., & Xu, Y. (2024). Lalonde (1986) after nearly four decades: Lessons learned. *arXiv preprint arXiv:2406.00827.*

Jin, H., & Rubin, D. B. (2008). Principal stratification for causal inference with extended partial compliance. *Journal of the American Statistical Association, 103*(481), 101–111.

M. Nguyen, *Experimental Design*, https://doi.org/10.1007/978-3-032-01839-7

Jin, H., & Rubin, D. B. (2009). Public schools versus private schools: Causal inference with partial compliance. *Journal of Educational and Behavioral Statistics, 34*(1), 24–45.

Johansson, P., & Schultzberg, M. (2022). Rerandomization: A complement or substitute for stratification in randomized experiments? *Journal of Statistical Planning and Inference, 218*, 43–58.

Kapelner, A., & Krieger, A. (2014). Matching on-the-fly: Sequential allocation with higher power and efficiency. *Biometrics, 70*(2), 378–388.

Keele, L., & Grieve, R. (2025). So many choices: A guide to selecting among methods to adjust for observed confounders. *Statistics in Medicine, 44*(5), e10336.

LaLonde, R. J. (1986). Evaluating the econometric evaluations of training programs with experimental data. *The American Economic Review, 76*, 604–620.

Lee, D. S. (2009). Training, wages, and sample selection: Estimating sharp bounds on treatment effects. *The Review of Economic Studies, 76*, 1071–1102.

Long, Q., Little, R. J., & Lin, X. (2010). Estimating causal effects in trials involving multitreatment arms subject to non-compliance: a bayesian framework. *Journal of the Royal Statistical Society: Series C (Applied Statistics), 59*(3), 513–531.

Mardia, K. V. (1970). Measures of multivariate skewness and kurtosis with applications. *Biometrika, 57*(3), 519–530.

Morgan, K. L., & Rubin, D. B. (2012). Rerandomization to improve covariate balance in experiments. *Annals of Statistics, 40*(2), 1263–1282.

Pearl, J. (2014). Graphical models for probabilistic and causal reasoning. *Computing handbook* (Vol. 1, 3rd ed., pp. 44–1).

Pearl, J. (2019). The seven tools of causal inference, with reflections on machine learning. *Communications of the ACM, 62*(3), 54–60.

Pearl, J., & Mackenzie, D. (2018). *The book of why: The new science of cause and effect.* Basic Books.

Pocock, S. J., & Simon, R. (1975). Sequential treatment assignment with balancing for prognostic factors in the controlled clinical trial. *Biometrics, 31*, 103–115.

Rosenbaum, P. R. (2002). Attributing effects to treatment in matched observational studies. *Journal of the American statistical Association, 97*(457), 183–192.

Rosenbaum, P. R., & Rubin, D. B. (1983). The central role of the propensity score in observational studies for causal effects. *Biometrika, 70*(1), 41–55.

Rubin, D. B. (1973). The use of matched sampling and regression adjustment to remove bias in observational studies. *Biometrics, 29*, 185–203.

Tukey, J. W. (1993). Tightening the clinical trial. *Controlled Clinical Trials, 14*(4), 266–285.

Wright, S. (1920). The relative importance of heredity and environment in determining the piebald pattern of guinea-pigs. *Proceedings of the National Academy of Sciences, 6*(6), 320–332.

Yule, G. U. (1899). An investigation into the causes of changes in pauperism in england, chiefly during the last two intercensal decades (part I.). *Journal of the Royal Statistical Society, 62*(2), 249–295.

Zhao, J., & Zhou, Z. (2025). Pigeonhole design: Balancing sequential experiments from an online matching perspective. *Management Science, 71*(3), 1889–1908.

MIX
Papier aus verantwortungsvollen Quellen
Paper from responsible sources
FSC® C105338

If you have any concerns about our products,
you can contact us on
ProductSafety@springernature.com

In case Publisher is established outside the EU,
the EU authorized representative is:
Springer Nature Customer Service Center GmbH
Europaplatz 3, 69115 Heidelberg, Germany

Printed by Libri Plureos GmbH
in Hamburg, Germany